全国高等中医药院校中药学类专业双语规划教材
Bilingual Planned Textbooks for Chinese Materia Medica Majors in TCM Colleges and Universities

药用植物学实验

Pharmaceutical Botany Experiment

（供中药学类、药学类及相关专业使用）

(For Chinese Materia Medica, Pharmacy and other related majors)

主　编　高德民　王光志

副主编　林　莺　刘湘丹　严玉平　陆　叶　包华音

编　者　（以姓氏笔画为序）

王光志（成都中医药大学）　　　方清影（安徽中医药大学）

包华音（山东中医药大学）　　　任　艳（西南民族大学）

任广喜（北京中医药大学）　　　刘　钊（河北中医学院）

刘湘丹（湖南中医药大学）　　　孙立彦（山东第一医科大学）

严玉平（河北中医学院）　　　　李先宽（天津中医药大学）

李思蒙（南京中医药大学）　　　吴廷娟（河南中医药大学）

吴清华（成都中医药大学）　　　吴靳荣（上海中医药大学）

陆　叶（苏州大学）　　　　　　陈　莹（陕西中医药大学）

邵艳华（广东药科大学）　　　　林　莺（滨州医学院）

秦　琴（成都医学院）　　　　　倪梁红（上海中医药大学）

高德民（山东中医药大学）　　　崔治家（甘肃中医药大学）

董丽华（江西中医药大学）　　　程丹丹［齐鲁工业大学（山东省科学院）］

樊锐锋（黑龙江中医药大学）

中国健康传媒集团
中国医药科技出版社

内 容 提 要

本教材是"全国高等中医药院校中药学类专业双语规划教材"之一，依据教育部相关文件和精神以及本专业的教学要求和实验课程特点编写而成。本教材分为总论和实验两部分内容。总论主要包括基本知识和实验技术。实验部分共安排 19 个实验，内容包括药用植物形态学、分类学和解剖学三部分。附录介绍了实验中常用试剂的配制方法，以方便学生查阅。

本教材实用性强，主要供全国高等中医药院校中药学类、药学类及相关专业的师生、研究人员使用。

图书在版编目（CIP）数据

药用植物学实验：汉英对照 / 高德民，王光志主编 . —北京：中国医药科技出版社，2020.7

全国高等中医药院校中药学类专业双语规划教材

ISBN 978-7-5214-1887-3

Ⅰ.①药…　Ⅱ.①高…②王…　Ⅲ.①药用植物学 – 实验 – 双语教学 – 中医学院 – 教材 – 汉、英

Ⅳ.① Q949.95-33

中国版本图书馆 CIP 数据核字（2020）第 102363 号

美术编辑　陈君杞

版式设计　辰轩文化

出版　**中国健康传媒集团** | 中国医药科技出版社

地址　北京市海淀区文慧园北路甲 22 号

邮编　100082

电话　发行：010-62227427　邮购：010-62236938

网址　www.cmstp.com

规格　889×1194 mm $\frac{1}{16}$

印张　8½

字数　221 千字

版次　2020 年 8 月第 1 版

印次　2023 年 12 月第 2 次印刷

印刷　三河市万龙印装有限公司

经销　全国各地新华书店

书号　ISBN 978-7-5214-1887-3

定价　**39.00** 元

获取新书信息、投稿、为图书纠错，请扫码联系我们。

近些年随着世界范围的中医药热潮的涌动，来中国学习中医药学的留学生逐年增多，走出国门的中医药学人才也在增加。为了适应中医药国际交流与合作的需要，加快中医药国际化进程，提高来中国留学生和国际班学生的教学质量，满足双语教学的需要和中医药对外交流需求，培养优秀的国际化中医药人才，进一步推动中医药国际化进程，根据教育部、国家中医药管理局、国家药品监督管理局等部门的有关精神，在本套教材建设指导委员会主任委员成都中医药大学彭成教授等专家的指导和顶层设计下，中国医药科技出版社组织全国50余所高等中医药院校及附属医疗机构约420名专家、教师精心编撰了全国高等中医药院校中药学类专业双语规划教材，该套教材即将付梓出版。

本套教材共计23门，主要供全国高等中医药院校中药学类专业教学使用。本套教材定位清晰、特色鲜明，主要体现在以下方面。

一、立足双语教学实际，培养复合应用型人才

本套教材以高校双语教学课程建设要求为依据，以满足国内医药院校开展留学生教学和双语教学的需求为目标，突出中医药文化特色鲜明、中医药专业术语规范的特点，注重培养中医药技能、反映中医药传承和现代研究成果，旨在优化教育质量，培养优秀的国际化中医药人才，推进中医药对外交流。

本套教材建设围绕目前中医药院校本科教育教学改革方向对教材体系进行科学规划、合理设计，坚持以培养创新型和复合型人才为宗旨，以社会需求为导向，以培养适应中药开发、利用、管理、服务等各个领域需求的高素质应用型人才为目标的教材建设思路与原则。

二、遵循教材编写规律，整体优化，紧跟学科发展步伐

本套教材的编写遵循"三基、五性、三特定"的教材编写规律；以"必需、够用"为度；坚持与时俱进，注意吸收新技术和新方法，适当拓展知识面，为学生后续发展奠定必要的基础。实验教材密切结合主干教材内容，体现理实一体，注重培养学生实践技能训练的同时，按照教育部相关精神，增加设计性实验部分，以现实问题作为驱动力来培养学生自主获取和应用新知识的能力，从而培养学生独立思考能力、实验设计能力、实践操作能力和可持续发展能力，满足培养应用型和复合型人才的要求。强调全套教材内容的整体优化，并注重不同教材内容的联系与衔接，避免遗漏和不必要的交叉重复。

三、对接职业资格考试，"教考""理实"密切融合

本套教材的内容和结构设计紧密对接国家执业中药师职业资格考试大纲要求，实现教学与考试、理论与实践的密切融合，并且在教材编写过程中，吸收具有丰富实践经验的企业人员参与教材的编写，确保教材的内容密切结合应用，更加体现高等教育的实践性和开放性，为学生参加考试和实践工作打下坚实基础。

四、创新教材呈现形式，书网融合，使教与学更便捷更轻松

全套教材为书网融合教材，即纸质教材与数字教材、配套教学资源、题库系统、数字化教学服务有机融合。通过"一书一码"的强关联，为读者提供全免费增值服务。按教材封底的提示激活教材后，读者可通过 PC、手机阅读电子教材和配套课程资源（PPT、微课、视频等），并可在线进行同步练习，实时收到答案反馈和解析。同时，读者也可以直接扫描书中二维码，阅读与教材内容关联的课程资源，从而丰富学习体验，使学习更便捷。教师可通过 PC 在线创建课程，与学生互动，开展在线课程内容定制、布置和批改作业、在线组织考试、讨论与答疑等教学活动，学生通过 PC、手机均可实现在线作业、在线考试，提升学习效率，使教与学更轻松。此外，平台尚有数据分析、教学诊断等功能，可为教学研究与管理提供技术和数据支撑。需要特殊说明的是，有些专业基础课程，例如《药理学》等 9 种教材，起源于西方医学，因篇幅所限，在本次双语教材建设中纸质教材以英语为主，仅将专业词汇对照了中文翻译，同时在中国医药科技出版社数字平台"医药大学堂"上配套了中文电子教材供学生学习参考。

编写出版本套高质量教材，得到了全国知名专家的精心指导和各有关院校领导与编者的大力支持，在此一并表示衷心感谢。希望广大师生在教学中积极使用本套教材和提出宝贵意见，以便修订完善，共同打造精品教材，为促进我国高等中医药院校中药学类专业教育教学改革和人才培养做出积极贡献。

全国高等中医药院校中药学类专业双语规划教材
建设指导委员会

数字化教材编委会

前　言

本教材是"全国高等中医药院校中药学类专业双语规划教材"之一。本教材的编写以党的二十大精神为指导，并根据本套教材的编写原则以及药用植物学实验教学大纲编写而成。药用植物学是中药学专业的基础课，实验教学在专业学习中占有重要的地位。药用植物学实验双语教学，不仅能加强学生基本技能的训练、增进国际学习和交流能力，也有利于中医药国际化发展。为此，本教材为中英文双语对照，英文专业词汇准确、易懂，以便于学生学习和交流。

本教材力求简洁，突出实践特色，体现系统性和代表性。本教材分为总论和实验部分。总论主要包括基本知识和实验技术。实验部分安排19个实验，内容包括药用植物形态学、分类学和解剖学三部分。附录介绍了实验中常用试剂的配制方法。各高校可根据实验学时数和内容等具体情况进行合并或调整。鉴于实验中分类部分材料受环境、季节的影响较大，教师可适当地灵活安排。每个实验后附有一定数量的实验报告题目和较为灵活的思考题，以便于学生发挥自己的创造力。

本教材具体编写分工如下：总论由程丹丹、倪梁红、陈莹、林鸢、孙立彦、吴靳荣、刘钊、任广喜、董丽华、秦琴编写。实验一由包华音编写，实验二由吴靳荣编写，实验三由陈莹编写，实验四由林鸢编写，实验五由董丽华编写，实验六由程丹丹、倪梁红、方清影、任广喜、刘钊编写，实验七由严玉平编写，实验八由孙立彦编写，实验九由崔治家编写，实验十由刘湘丹编写，实验十一由邵艳华编写，实验十二由吴廷娟编写，实验十三由李思蒙编写，实验十四由李先宽编写，实验十五由包华音编写，实验十六由任艳编写，实验十七由陆叶编写，实验十八由吴清华编写，实验十九由樊锐锋编写。附录一由方清影编写，附录二由包华音编写。全书由高德民、王光志负责统稿、定稿。本教材编写得到了所有编者及其所在单位领导的大力支持与帮助，在此一并表示衷心感谢！

由于受编者学识所限，难免存在疏漏与不足之处，恳请广大读者批评指正，以便日臻完善。

编　者

Preface

    The textbook was compiled according to the basic principles of compiling bilingual teaching materials for pharmacy in Chinese medicine colleges and universities and the outline of compiling experiments in medicinal botany, which is approved by the textbook compilation and review committee. Medicinal botany is the basic course of traditional Chinese medicine (TCM), and experimental teaching plays an important role in the professional learning. Carrying out bilingual teaching of medicinal botany experiment can not only strengthen the training of students' basic skills, enhance their international learning and communication ability, but also contribute to the internationalization of traditional Chinese medicine. So this textbook is equipped with Chinese and English. English professional vocabulary is accurate and easy to understand, so as to facilitate students' learning and communication.

    The textbook strives for simplicity, highlights the practical features, and embodies systematicness and representativeness. The textbook includes an general instructions and experiments. The general instructions mainly includes basic knowledge and experimental techniques. In the experimental part, 19 experiments were arranged, including morphology, taxonomy and anatomy of medicinal plants. The appendix introduces the preparation methods of common reagents used in experiments. Colleges and universities may flexibly adjust experimental contents. Each experiment is accompanied by a certain number of experimental report questions and more flexible thinking questions, so that students can give play to their creativity.

    The specific division of work was as follows. The general instruction was compiled by Cheng Dandan, Ni Lianghong, Chen Ying, Lin Ying, Sun Liyan, Wu Jinrong, Liu Zhao, Ren Guangxi, Dong Lihua, Qin Qin. Experiment 1 was written by Bao Huayin, experiment 2 by Wu Jinrong, experiment 3 by Chen Ying, experiment 4 by Lin Ying, experiment 5 by Dong Lihua, experiment 6 by Cheng Dandan, Ni Lianghong, Fang Qingying, Ren Guangxi and Liu Zhao, and experiment 7 by Yan Yuping. Experiment 8 by Sun Liyan, experiment 9 by Cui Zhijia, experiment 10 by Liu Xiangdan, experiment11 by Shao Yanhua, experiment 12 by Wu Tingjuan, experiment 13 by Li Simeng, experiment 14 by Li Xiankuan, experiment 15 by Bao Huayin, experiment 16 by Ren Yan, experiment 17 by Lu Ye, experimental 18 by Wu Qinghua, experiment 19 by Fan Ruifeng. Appendix 1 was written by Fang Qingying, and appendix 2 by Bao huayin. The textbook was reviewed and revised by Gao Demin and Wang Guangzhi. The compilication of this textbook hed received the strong supports and help from all the editors. Thank you very much.

    Inevitably there are omissions and deficiencies in the book, we especially invite readers to criticize and correct.

目录 | Contents

总 论
Introduction

实 验 部 分
Experimental Section

总 论

一、显微镜的使用

（一）显微镜的构造

根据所用光源的不同，显微镜可分为光学显微镜和电子显微镜两大类。光学显微镜的有效放大倍数可达 1250 倍，最高分辨率为 0.2μm，而电子显微镜是放大倍数可达 80 万~120 万倍，其分辨率比光学显微镜大 1000 倍。

最常用的为复式显微镜，有单筒镜和双筒镜两种类型，它们的基本构造包括光学系统和机械系统（图 1）

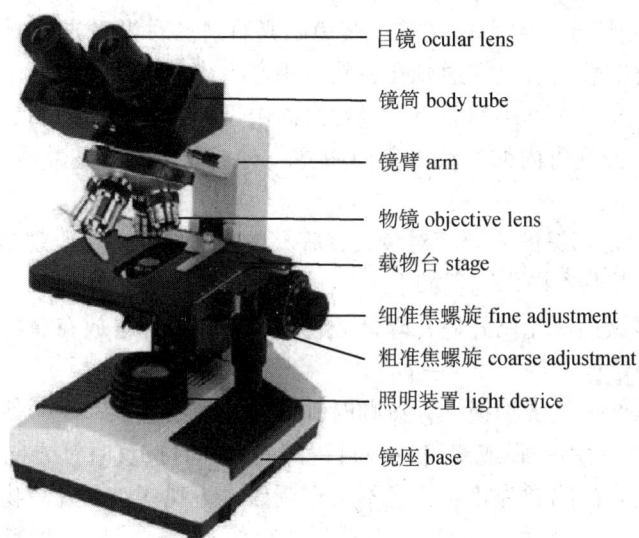

目镜 ocular lens

镜筒 body tube

镜臂 arm

物镜 objective lens

载物台 stage

细准焦螺旋 fine adjustment

粗准焦螺旋 coarse adjustment

照明装置 light device

镜座 base

图 1 复式显微镜

Fig.1 Compound microscope

1. 光学系统

（1）物镜 安装在镜筒下端的物镜转换器上，可分为低倍、高倍和油浸物镜三种。

（2）目镜 安装在镜筒上端，有单目镜和双目镜两种类型，可将物镜所成的像进一步放大，常用目镜的放大倍数有 8×、10×、15× 等。放大倍数越低其镜头的长度越长。双筒复式显微镜的目镜可以调焦，提高了成像质量。

（3）反光镜 是由平面镜和凹面镜组成的两面镜。兼有反光和聚集光线的作用，将光线反射在聚光器上。

（4）聚光器　装于载物台下，由聚光镜和虹彩光圈等组成，它可将平行的光线汇集成束，集中于一点以增强被检物体的照明，也可上下调节，来控制光度。

2. 机械系统

（1）镜座　显微镜的底座，用以稳固和支持整个镜体。

（2）镜柱　镜座上面直立的短柱，支持镜体上部的各部分。

（3）镜臂　下连镜柱，上连镜筒，为取放镜体时手握的部分。

（4）镜筒　显微镜上部圆形中空的长筒，其上端安放目镜，下端与物镜转换器相连，并使目镜和物镜的配合保持一定距离。镜筒能保护成像的光路和亮度。

（5）物镜转换器　为接于镜筒下端的圆盘，盘上有3~4个安装物镜的螺旋口。当旋转物镜转换器时，物镜即可固定在使用的位置上，保证物镜与目镜的光轴同心。

（6）载物台　为放置玻片标本的平台，中央有一通光孔。上面有玻片卡尺和一个推进器，用以固定玻片标本，同时可以向前后左右各方向移动。

（7）调焦装置　用以调节物镜和标本之间的距离，以便得到清晰的物像。在镜柱两侧有粗、细调节螺旋各一对，旋转时可使载物台上升或下降。大的一对为粗准焦螺旋，小的一对为细准焦螺旋。

（二）显微镜的使用方法

1. 取镜和放置　按固定编号从镜柜里取出显微镜。取镜时右手握住镜壁，左手平托镜座，保持镜体直立，轻轻放置于座位的左前方，距桌边约10cm处，以便于观察和防止掉落。

2. 对光　常用日光灯或自然光作光源，避免阳光直射。对光时先从低倍物镜开始，即先把低倍物镜转到中央，对准载物台上的通光孔，然后用左眼或双眼从目镜向下注视，同时转动反光镜，使镜面向着光源，光弱时可用凹面镜。当在镜筒内见到一个圆形而明亮的视野时，再利用虹彩光圈调节光的强度，使视野内的光线均匀而明亮。对于电光源的显微镜，直接打开光源开关，然后调节光的强度即可。

3. 低倍镜的使用　低倍镜的视野相对较大，容易发现目标和确定要观察的部位。所以观察标本时，应先用低倍镜。其基本操作过程如下。

（1）放置切片　显微镜对光调好后，降低载物台，把玻片标本放在载物台中央，使标本正对通光孔，然后用卡尺固定标本片。

（2）调整焦点　从侧面注视物镜，并按顺时针方向转动粗准焦螺旋，使载物台徐徐上升至物镜离玻片约5mm处。然后用左眼或双目注视目镜筒内，同时按反时针方向转动粗准焦螺旋使载物台下降，直到看见清晰的物像为止（注意不可在调焦时边观察边上升载物台，否则会使物镜和玻片触碰，压碎玻片，损伤物镜）。如果一次看不清物像，应重新检查材料是否放在光轴上，重新移正材料，再重复上述操作过程直至物像出现和清晰为止。

（3）低倍镜的观察　焦距调好后，可根据需要前后左右移动玻片使要观察的部分在最佳位置上。找到物像后，可根据标本的厚薄、颜色、成像反差强弱是否合适等调节光线强度。

4. 高倍镜的使用　从低倍镜到高倍镜应注意以下几点。

（1）在使用高倍镜前应在低倍镜中选好目标并将其移至视野中央，然后转动物镜转换器，再转换成高倍物镜。

（2）转换成高倍物镜之后，在视野中即可见模糊物像，只要稍调动细准焦螺旋，即可见到清晰的物像。

（3）在换用高倍显微镜观察后，视野变小变暗，需重新调节视野的亮度。

5. 油镜的使用　先在低倍镜下找到被检部分，再换成高倍镜调整焦点，并将被检部位移至视野中央，然后再换成油镜。使用油镜时，要先在盖玻片上滴加一滴香柏油，才能使用。油镜使用后，应立即以擦镜纸蘸少许二甲苯擦去镜头上的油迹。

6. 收镜　观察完毕，应先降低载物台，取下玻片，转动物镜转换器，使物镜镜头与通光孔错开，再升高载物台，并将反光镜还原成与桌面垂直，擦净镜体，罩上防尘罩。仍用右手握住镜臂，左手平托镜体，放回原处。

（三）显微镜的保养

1. 必须熟悉并严格执行显微镜的使用步骤和要求。

2. 保持清洁，机械部分可用细软布擦拭。光学部位要用擦镜纸轻擦，切忌用手或其他粗糙物擦拭，以免损伤镜面。

3. 在桌面上移动显微镜时，应轻拿轻放，切忌拖拉震动显微镜。

4. 物镜或目镜不可任意摘下或打开，以免灰尘落入。

5. 观察的标本必须盖盖玻片，带水或试剂的玻片标本，必须两面擦干再放载物台上观察。

6. 显微镜为精密仪器，不能任意拆修，如遇机件失灵，使用困难时，不可强行转动，应立即报告指导老师解决，以免造成损坏。

7. 显微镜不用时应盖上防尘罩，及时放回镜柜内，不可放于日光下暴晒，不可与化学试剂放在一起。镜柜内应放干燥剂防潮。

习题

Introduction

I Operation and Maintenance of Microscope

1. The structure of microscope

According to the difference of optical resource, microscopes can be divided into optical microscopes and electron microscopes. The effective magnification of optical microscope can reach 1250 times, and the highest resolution is $0.2\mu m$, while the magnification of electronic microscope can reach up to $0.8\sim1.2$ million, 1000 times larger than a optical microscope.

The most commonly used is the compound microscope, which has one drawtube or two drawtubes. Both of the two kinds above have optical systems and mechanical systems (Fig.1).

1.1 Optics system

1.1.1 Objective lens   It is settled on the revolving nosepiece and can be classified as low-power, high-power and oil-immersion.

1.1.2 Ocular lens   It is settled on the top of the lens tube and can magnify the image from the objective. It has magnitude times on it such as $8\times$, $10\times$ and $15\times$. The lower of the magnification, the longer of the lens length. Ocular has two type: one-drawtube ocular and two-drawtube ocular. The latter can be adjusted to improve the quality of image.

1.1.3 Mirror   The two-sided mirror is composed of a plane mirror and a concave mirror, which has the function of reflecting and concentrating light. The mirror can reflect the light on the condenser adjustment.

1.1.4 Condenser   It is settled under the stage and is made up of a condenser mirror and an iris diaphragm. It can collect the parallel light into a beam to strengthen the illumination on the specimen, and it can also be adjusted up and down to control the luminosity.

1.2 Mechanical systems

1.2.1 Base   The foundation of the microscope upholds the whole apparatus and makes it steady.

1.2.2 Column   An erect short column on the base upholds the department above it.

1.2.3 Arm   The arm is connected with the base and column, which is the part held by the hand when the microscope is taken.

1.2.4 Body tube   This is a long tube to separate the objective lens and the ocular, which can protect the light way and brightness of the image.

1.2.5 Revolving nosepiece   A disk under the column can be turned easily. There are 3-4 screw poles on it. When it is turned, the objective lens can be fixed at a certain place, and make the light of the objective lens and the ocular lens on a straight line.

1.2.6 Stage   On which the specimen is placed. There is a core in the central. It has a caliper and a thruster on it, which can fix the slide and move on each position.

1.2.7 Adjustment   It is used to adjust the distance between the objective lens and the specimen so as to get a clear image. On both sides of the column, there are a pair of coarse and fine adjustments, which can make the stage rise or fall when rotating. The large pair is coarse adjustments, and the small pair is fine adjustments.

2. The use of microscope

2.1 Fetch and display   Get out the microscope from the cupboard with the right number. You should hold the arm with your right hand and support the base with your left hand to make the whole body steady. The microscope should be put on the left side of your table with a distance from the table edge about 10cm in order to facilitate observation and prevent it from falling down.

2.2 Adjustment of the light   We use the light shining from the window or the light from daylight lamp, not direct shining sunlight. Turn the low power lens to the central right onto the pore and then observe the specimen with your left eye or both. Meanwhile, turn the mirror to make the surface to the light. You may use the concave mirror when the light is weak. Once you see the bright round vision inside, turn the iris diaphragm for better light power to make the vision clear. For the electron microscope, just to turn on the power and intensity of light.

2.3 The use of low-power objective lens   The low-power objective lens can give a wider vision field, and you can find the destination and assure the exact part of the objective you want easily.

2.3.1 Lay the slice up   After adjustment of the light, lower the stage, set the specimen right on the central of the stage and align the specimen with the optical aperture, and then fixed the two sides of the slice with the remover.

2.3.2 Adjust the focus   Look at the objective lens with your eyes from the side, slowly turn the coarse adjustments clockwise to raise the stage until the distance between the stage and the slide is about 5mm. Observe the vision with your left eye or both of your eyes and turn the coarse adjustments counterclockwise until you see the image precisely (Don't lift up the stage while you observe, otherwise you may smash the objective lens and the slide, break the lens or the glasses). If you can't see the image clearly, you should check whether the slide is placed on the optical axis, reposition the material, and repeat the above process until the slide becomes clear.

2.3.3 Observe with the low-power objective lens   After the focus is adjusted, you can move the slide to the best place as you want to. After the object image is found, the light intensity can be adjusted according to the thickness, color of the specimen, and the intensity of imaging contrast.

2.4 The use of high-power objective lens   The following points should be noted from low to high power objective lens.

2.4.1 Before using the high-power objective lens, you should make your destination right in the central of the field of vision with the low-power objective lens. Then turn to the high-power lens.

2.4.2 After being converted into a high-power objective lens, the blurred object image can be seen in the field of vision, and the clear object image can be seen by slightly adjusting the fine adjustment.

2.4.3 After changing to high power microscope, the visual field becomes smaller and darker, and the brightness of the visual field needs to be adjusted again.

2.5 Use of the oil-immersion   Before using the oil-immersion, find the part you want with the low-power objective, adjust the focus with the high-power one, take the part to the central field of vision, and then turn it to the oil-immersion. Drip a drop of cedar oil on the top glass before using the oil immersion. After using the oil-immersion, tip it with a few drops of xylene immediately to wipe off the oil spot.

2.6　Put back the microscope　After using the microscope, lower the stage, remove the glass slide, rotate the resolving nosepiece to stagger the objective lens with the aperture, then raise the stage, and restore the reflector to be perpendicular to the desktop, wipe the mirror body, cover the dustproof cover. Still hold the mirror arm in your right hand and the mirror body in your left hand, put microscope back to the place in the same way you brought it out.

3. The maintenance of microscope

3.1　Be familiar with and strictly follow the procedures and requirements for the use of microscope.

3.2　Keep the microscope clean. Machine parts can be wiped with a soft cloth. Optical parts to be brushed with lens wiping paper, avoid hand or other rough things to wipe, so as not to damage the lens.

3.3　When moving the microscope on the desktop, it should be gently handled, never drag and shake microscope roughly.

3.4　The objective lens or the ocularlens should not be arbitrarily removed or opened, so as the dust not to fall into it.

3.5　The specimen must be covered with glass slides, and the excessive water or reagents must be dried on both sides before being placed on the stage for observation.

3.6　The microscope is a precise instrument, and do not be arbitrarily disassembled and repaired. If there are something out of control about the microscope, do not be forcibly turned, you should immediately report to the instructor to solve, so as to avoid damage.

3.7　When the microscope is not in use, it should be covered with dust cover and put back into the mirror cabinet in time. It should not be exposed to the sun or put together with chemical reagents. Desiccant should be placed in the mirror cabinet to prevent moisture.

二、基本实验技术

（一）徒手切片法

徒手切片法是简便、快捷的一种切片方法，具体步骤如下。

1. 材料准备 一般草本植物的根、茎、叶等，先切成3cm左右的段块，断面不超过 3~5mm²。较小、软或薄的材料，如叶片，可用胡萝卜根、马铃薯块茎等作为支持物，夹住材料一起切片。较坚硬的材料，如木材等，可用水煮等方法进行软化后再切片。

2. 切片 用左手的拇指、食指和中指夹住材料，并使材料高于手指之上；右手执刀片，平放于左手食指之上，刀口向内，与材料断面平行；左手固定不动，右臂移动用力，从左前方往右后方滑行切片，切片过程中可用水润湿刀面和材料切面，如此连续操作，切得薄片。

3. 选片 将薄片用湿毛笔从刀片上轻轻移到盛水的培养皿中。用毛笔挑选薄且完整的切片，放在载玻片上，制成临时装片观察。

（二）临时装片法

临时装片法是用少量材料，如薄的表皮、切成的薄片或粉末等，置于载玻片上的水滴中，加盖盖玻片制成玻片标本，或选用甘油醋酸试液、水合氯醛试液处理后观察。

加盖玻片时应注意先用镊子轻轻夹住盖玻片，使其边缘与材料边缘水滴的接触，然后慢慢放下，放平盖玻片，使盖玻片下的空气逐渐被水挤出而不产生气泡，以免影响观察。

（三）滑走切片法

滑走切片法是用滑走切片机进行切片，制片方法与徒手切片类似。滑走切片可以根据需要调节切片厚度，切制出厚薄均匀完整的切片，但不能进行连续操作。此法适合切制木材或较硬的组织材料。

（四）解离组织法

解离组织法是利用化学试剂溶解细胞的胞间层而使细胞彼此分离的一种方法，以便观察不同组织的细胞特征。该方法常用于观察木质化程度较高的组织，如纤维、石细胞、导管及管胞等。在进行组织解离前，应先将材料切成宽或厚约2mm的小条或片。常用的解离方法有硝铬酸解离法、氯酸钾和硝酸解离法和氢氧化钾解离法。

1. 硝铬酸解离法 置材料于试管中，加入适量10%硝酸与10%铬酸的等量混合液，室温放置24h左右，或加温（30~45℃）以缩短解离时间，至用玻璃棒挤压能离散为止，倾去酸液，加水洗涤干净，取少许材料置载玻片上，用解剖针离散，稀甘油装片后观察。

2. 氯酸钾和硝酸解离法 置材料于试管中，加入50%硝酸溶液及少量氯酸钾，缓缓加热，待产生的气泡渐少时，再及时加入少量氯酸钾，以维持气泡稳定的产生，至用玻璃棒挤压能离散为止，倾去酸液，加水洗涤干净，照硝铬酸解离法操作装片观察。此法和硝铬酸解离法适用于质地较坚硬、木化组织多的材料。

3. 氢氧化钾解离法 置材料于试管中，加入适量 5% 氢氧化钾溶液，加热至用玻璃棒挤压能离散为止，倾去碱液，加水洗涤干净，照硝酸银解离法操作装片观察。此法适用于含薄壁组织多、木化组织少的材料。

（五）石蜡切片法

石蜡制片是观察植物组织构造常用的制片方法之一。该方法应用石蜡与植物组织能够很好地结合这一基本原理，经过一系列方法制成透明的薄片，并可长期保存。但此法所需操作时间长，步骤复杂。具体步骤如下。

1. 取材 取新鲜或干燥材料切成大小适中的小块。

2. 固定或软化 材料放入装有固定液的试剂瓶中浸泡 12~24h。固定液常用 F.A.A 固定液。干燥材料用水或甘油浸软后取材，不需固定。

3. 冲洗 固定后用 50% 乙醇冲洗数次。

4. 脱水 固定好的材料浸泡至不同浓度的乙醇中脱水。乙醇浓度梯度通常依次为 50%、70%、80%、90% 和 100%。每个梯度浸泡时间 1~3h，视不同材料而定。

5. 透明 用 50% 的二甲苯乙醇溶液浸泡 2~3h 后，用纯二甲苯浸泡两次，每次 1~2h。

6. 浸蜡 一般室温下先用石蜡和二甲苯的混合液浸泡已透明的材料。浸泡时间一般可为 1~2天。将固体纯石蜡放置于恒温箱内融化，恒温箱温度一般设置为高于石蜡熔点 2~3℃。将材料从石蜡和二甲苯的混合液中转移至融化后的石蜡中。每隔 2h 换一次纯石蜡，共换 3 次。浸蜡时间可根据材料质地、大小适当改变。

7. 包埋 材料和石蜡趁热一起倒入纸盒中。用加热的镊子调整好材料的位置后，降温使之凝固（可放入凉水中）。

8. 切片 取包埋好的蜡块修成上下两边平行的长方体。修好的蜡块固定在切片机上并被切成 5~10μm 厚的连续蜡带，用毛笔小心取下蜡带后备用。

9. 贴 在涂抹少许粘片剂的干净载玻片上滴加数滴蒸馏水。将切好的蜡带用镊子转移至蒸馏水中。常用的粘片剂为郝伯特（Haupt）粘贴剂或蛋清，粘片剂中可加入水杨酸钠等防腐剂以提高保存时间。

10. 展片 把载玻片放置于 40℃展片台上，蜡片受热而伸展摊平。展片后把载玻片转移到 37℃的恒温箱内烘干 12h 以上。

11. 脱蜡 将粘有蜡片的载玻片浸于纯二甲苯中，约 10min，使石蜡完全溶去，以便染色。

12. 染色 切片依次移入 50% 的二甲苯纯乙醇溶液、95%、85%、70%、50%、30% 乙醇、纯水中，每次时间约为 5~10min。然后浸入 1% 番红水溶液 1~24h 染色。取出水洗后，依次用 30%、50%、70% 乙醇脱水。浸入 0.5% 固绿乙醇溶液 10~40s 染色。

13. 封片 染色后的切片经梯度乙醇脱水、二甲苯透明后滴加适量封藏剂，再加干净的盖玻片倾斜放下，进行封片。封藏剂常用加拿大树胶或者中性树胶。

（六）生物绘图技术

绘图是药用植物学实验报告的重要组成部分之一，务必要保持对于事物描绘的客观性和准确性，应做到主次分明、画面统一、生动活泼，辅以一定的艺术性。绘图时，应避免涂抹，用圆点的疏密和衬影线条表示植物器官、组织的明暗对比和质量差异。图注一律用水平直线在图的右侧引出标注，标注内容多时可用折线，但必须整齐一致，所有图注需保持一列。不可用弧线、箭头线及交叉线做标注。图题和所用材料的名称及部位均写在图的下方，在绘制显微结构图时，需注明放大倍数。

1. 绘图工具 植物科学绘图包括形态图和结构图，一般所绘制的图为墨线图，所用工具较为简单。

（1）笔 实验报告中的墨线图，一般用铅笔绘制，常用 HB 中性铅笔（轮廓勾画清晰，便于修改），以及 2H-3H 硬性铅笔（便于描绘植物、药材等形态特征）。

（2）纸 实验报告常用无格的实验报告纸或白报本。

（3）其他 直尺、卡尺、比例尺、绘图橡皮等。

2. 药用植物和药材形态的绘图方法 药用植物和药材形态的绘图，应结合标本特点，利用墨线图的形式予以展现。从整体上把握标本的形态和各部分的特征，绘出轮廓，再在各部分中填绘细节特征。

在绘制植物时，通常将植物主体绘出，预留各器官的绘画空间，在根据各重要器官的特性进行详细的描绘。对于部分重要特征，可描绘放大图以展现其特征。

3. 显微结构图的绘图方法 显微结构图即为药用植物的构造图，利用显微镜对植物细胞、组织和器官的内部构造（横切、纵切、表面观）特征进行观察并进行绘图，包括绘制组织简图和组织详图。

（1）组织简图 组织简图是利用各类简易线条表示各种组织的界线，用各类特定的符号表示组织类型和特征的生物绘图法（图2），在绘图过程中，无需绘制植物细胞的形状。

（2）组织详图 组织详图是以细胞的详细形状、特征以及分布特点所绘制的细胞、组织、器官构造图。组织详图包括器官构造图、解离组织构造图、组织粉末构造图等，要求详尽地勾勒出器官、组织、细胞的内部特征，如细胞壁厚度、增厚的层纹、导管的类型、叶肉细胞类型、后含物类型等。在绘制器官构造时，可根据要求，绘制器官全图或局部图。

（3）绘制方法 将绘图所需的图纸置于显微镜右侧，左眼观察显微镜内的物象，右眼注视图纸完成绘图。在绘图时需选择特征较为明显的部分，首先用 2H 铅笔在绘图纸上勾勒出草图，再仔细观察标本内部，最后用 HB 铅笔绘制成图。

（七）显微化学鉴别

1. 显微化学鉴别的概念 显微化学鉴别法是将通过徒手切片法制成的植物材料，利用化学反应或染色的方法，使器官、组织或细胞内的某种化学成分（淀粉粒、蛋白质、脂肪等）在切片内予以显示，在显微镜下直接观察这些物质的形态、颜色和分布的一种快速定性及定位的方法。

2. 显微化学鉴别的适用范围 这种方法除广泛用于植物器官、组织和细胞后含物的研究外，也用于区分细胞壁的结构。

3. 显微化学鉴别的注意事项

（1）选用新鲜、健康的植物材料。

（2）取材后要尽快进行制片，以冰冻切片或徒手切片法制成薄片最为适宜。

（3）切片不宜太薄，否则材料中内含物太少导致化学反应现象不明显，材料的厚度应在 $20\sim40\mu m$ 较为适宜。

（4）选用的试剂和化学反应方法必须是对某一待测物质具有专属性。

（5）在取样的数量或观察、鉴别的次数上都要有重复，尽量排除实验误差。

（6）进行显微化学鉴别时，应注意设置对照试验。

微课

微课

表皮 epidermis	木栓层 phellem	厚角组织 collenchyma
针晶 acular crystal	方晶 solitary crystal	砂晶 micro-crystal
簇晶 cluster crystal	纤维 fiber	石细胞 sclereid
形成层 cambium	韧皮部 pliloem	木质部 xylem
射线 ray	裂隙 crack	分泌组织 secretory tissue

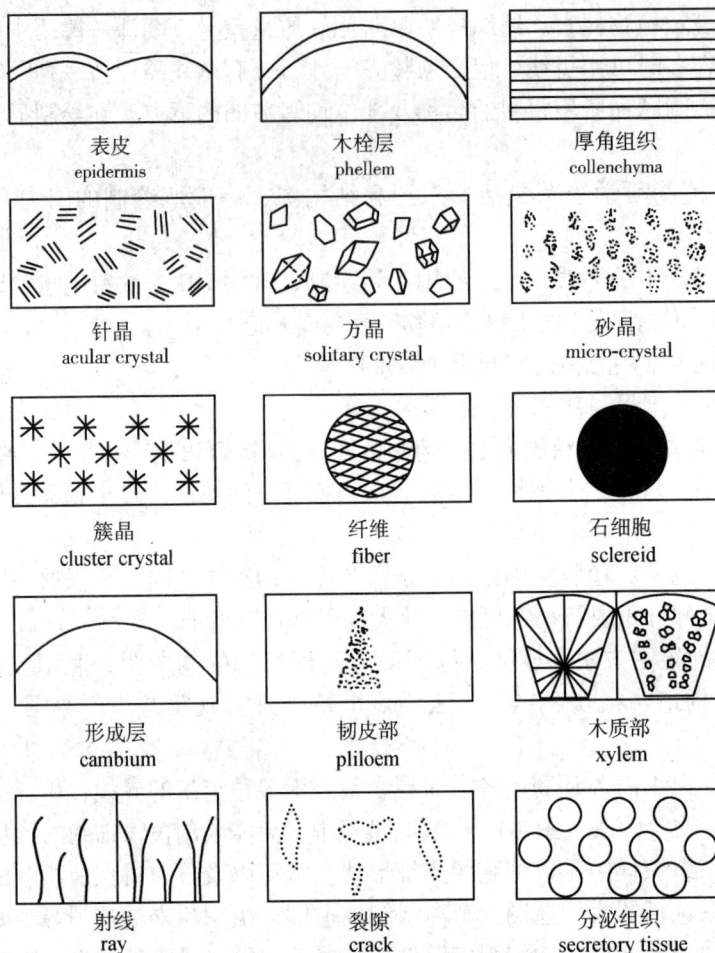

图2 植物组织简图常用表示方法

Figure 2 Metod of plant tissue diagram representation

（7）反应处理的切片和对照的切片应该尽量邻近，最好是前一片作反应处理切片，后一片则作对照切片。

（8）随时将观察的结果进行记录、拍照。

4. 常见的显微化学鉴别 这里主要介绍细胞壁化学成分和细胞后含物的显微化学鉴别方法。

（1）细胞后含物的显微化学鉴别

①淀粉的鉴定 淀粉是植物体中最主要的储藏物质，通常以淀粉粒的形式存在于植物细胞中，在不同植物的细胞中呈各种不同形状的颗粒。鉴定淀粉常用的方法是碘 - 碘化钾法。淀粉和碘反应形成碘化淀粉，呈现特殊的蓝色或棕红色反应。在实验中应该注意，碘液不宜过浓。

②菊糖的鉴别 菊糖由果糖分子聚合形成，主要存在于山茱萸科、菊科、龙胆科和桔梗科部分植物的细胞中。菊糖易溶于水，不溶于乙醇，为了能够观察到菊糖结晶，应将实验材料提前一周用95%的乙醇溶液浸泡，使菊糖晶体析出。然后向含有菊糖的材料滴加10% α-萘酚的乙醇溶液，再加80%硫酸1滴，显微镜下观察菊糖出现紫红色，但很快会溶解。

③蛋白质（糊粉粒）的鉴定 植物细胞内储存的蛋白质常呈固体颗粒——糊粉粒的形式存在，最常用的显微化学测定方法碘 - 碘化钾试剂法。该方法简单，可与鉴定淀粉同时进行，蛋白质被染成黄色，而淀粉则被染成蓝紫色。

④脂肪和脂肪油的鉴定 常用的显微化学鉴定方法是用苏丹Ⅲ或苏丹Ⅳ乙醇溶液染色，脂肪或脂肪油滴均被染成橘红色。但此法不能作为脂肪存在与否的证据，因树脂、挥发油、木栓化和

角质化的细胞壁也可以被苏丹Ⅲ或苏丹Ⅳ乙醇溶液染成红色，但脂肪经染色后更明显。

⑤晶体的鉴定　植物细胞含有的晶体以草酸钙结晶和碳酸钙结晶为常见，其中以草酸钙结晶最为常见。草酸钙晶体或碳酸钙晶体的形态特征以及存在的部位可作为中药材鉴定的依据之一。二者可用醋酸、稀盐酸或浓硫酸来鉴别，草酸钙结晶不溶于稀醋酸，加稀盐酸或浓硫酸溶解而无气泡产生，而碳酸钙结晶加醋酸和稀盐酸或浓硫酸则溶解，同时有 CO_2 气泡产生。

（2）细胞壁化学成分的显微化学鉴定

①纤维素的鉴定　细胞壁主要由纤维素构成，常用的鉴定方法是氯化锌 - 碘液法。将材料做徒手切片，取较薄的置于载玻片上，加 1~2 滴氯化锌 - 碘试液染色，显微镜下观察纤维素被染成蓝色或紫色。或者也可以用碘 - 碘化钾试液和 66% 浓硫酸染色效果也较好。

②木质化细胞壁的鉴定　木质化是指细胞壁内填充了木质素，可使细胞壁的硬度增强，细胞群的机械力增加。常将切成薄片的材料，滴加 1~2 滴浓盐酸后，再加间苯三酚试液。在显微镜下可以观察到木质化的细胞壁经过染色后呈现紫红色或樱桃红色。

③木栓化细胞壁的鉴定　细胞壁的木栓化是由于细胞壁中增加了脂肪性化合物——木栓质，对植物体内部组织具有保护作用。木栓化细胞壁常呈黄褐色，为死细胞。常用苏丹Ⅲ试液染色，在显微镜下可以观察到木栓化细胞壁被染成橘红色或红色。

④角质化细胞壁的鉴定　细胞中的原生质体产生的脂肪性化合物角质无色透明，它会在细胞壁内和表面堆积，使细胞壁内角质化，细胞壁外积聚形成角质层，二者都可以防止水分蒸发和微生物的侵害，起到保护植物内部组织的作用。角质化细胞的鉴定与木栓化相同，加入苏丹Ⅲ试液显橘红色或红色。

⑤黏液质化细胞壁的鉴定　黏液质化是细胞壁中所含有的果胶质和纤维素等成分变成黏液的一种变化。车前子、芥菜籽、亚麻子和丹参果实的表皮细胞中都具有黏液化细胞。常采用玫红酸钠乙醇溶液鉴定，显微镜下观察被染成玫瑰红色。

⑥矿质化细胞壁的鉴定　细胞壁的矿质化由细胞壁中增加硅质（二氧化硅或硅酸盐）或钙质导致。矿质化可以增强植物的机械支持能力。禾本科植物的茎、叶，木贼茎以及硅藻的矿质化细胞壁内含有大量的硅酸盐。通常加入硫酸或醋酸加以鉴定，硅质化细胞壁均不溶于二者，这种方法可用于鉴定硅质是否存在。

习题

Elementary Experiment Technology

1. Freehand section method

Freehand section is a convenient section method with short operating time. The method of freehand section is as follows.

1.1 Material preparation Cut the root, stem and leaf of herbaceous plants into segments of about 3cm, with a cross section no more than 3-5 mm^2. The small, soft, or thin material, such as leaves, could be cut with root of carrot or tuber of potato as support. Hard material, wood for example, should be softened by boiling and then sliced.

1.2 Section Clasp the material using the thumb, forefinger and middle finger of left hand and hold it above the fingers. Hold the razor blade with right hand and put it on the forefinger of left hand, with the edge of razor blade facing inward and parallel with the section of material. With the left hand fixed, the right arm is moved to move the razor blade from left-front to right rear. During the slicing process, the edge of razor blade and the section of material should keep wet. Do this continuously and get slice.

1.3 Choosing slice Remove the slices from the razor blade gently with a wet brush and place it in a basin full of water. Choose thin and complete slices with a brush and place them on a glass slide to make temporary slices for observation.

2. Method of making temporary section

The method of making temporary section is to place a little materials, such as a thin epidermis, thin slice or powder of samples into the water drop on the slide, then put the cover glass on the slide. You can also choose glycerine acetic acid solvent, chloral hydrate and other reagent to make temporary section.

When placing a cover glass, please pay attention to clamp the cover glass slightly with tweezers to make sure that its margin connects with the edge of water droplet on the edge of material, then put the cover glass down slowly to assure the air under the cover glass can be eliminated without air bubble, so as to avoid obstructing observation.

3. Method of sliding slice

Slide slice is a slicing method using a slide microtome and similar to that of freehand slicing. The advantage of sliding slice is that the thickness can be adjusted, and the microsection is integrity, while the operation isn't continuous. This method is suitable for cutting wood or hard tissue materials.

4. Method of dissociation tissue

The method of dissociation tissue is a method using chemical agents to dissolve the intercellular layer to make cells separate, so as to observe the cellular characteristics of different tissues. This method is often used for the tissue that is lignified strongly, such as fibers, stone cells, catheters and tracheids. Cut the samples into 2mm slices before dissociation. The commonly used dissociation methods are nitrochromic acid dissociation, potassium chlorate and nitric acid dissociation and potassium hydroxide dissociation.

4.1 Nitrochromic acid dissociation　Put the materials into tubes, add a mixed solution with the same volume of 10% nitric acid and 10% chromic acid, place about 24h at room temperature, or heat（30-45℃）to shorten the dissociation time, till its dissociation by compression with glass rod, drop out the acid solvents, wash clean with water, take a little material onto the slide, tear it with anatomical needle, treat it with dilute glycerin, cover it with the coverslip, then observe it under the microscope.

4.2 Potassium chlorate and nitric acid dissociation　Put the material into tubes, add some 50% nitric acid solution and a small amount of potassium chlorate, heat slowly. When the bubbles are gradually reduced, then add a small amount of potassium chlorate in time to maintain the stable generation of bubbles, till its dissociation by compression with glass rod. Drop out the acid solvent, wash clean with water, then treat it as 4.1 to make slide, then to observe. Methods 4.1 and 4.2 are suitable for the hard samples containing more xylem tissues.

4.3 Potassium hydroxide dissociation　Put the material into tubes, add an appropriate amount 5% potassium hydroxide, solution and heat until its dissociation by compression with glass rod, drop out the basic solvent, wash clean with water, then treat it as 4.1 to make slide, then to observe. This method is suitable for the samples containing many parenchyma and little xylem tissues.

5. Methods of paraffin section

Paraffin section is a common section method for observing the tissue structure of plant. The basic principle of this method is that paraffin can be well combined with plant tissue. The paraffin section is made through series of experimental procedures and stored for long periods of time. However, the required operation time is long and the steps are complicated. Paraffin section preparations are as follows.

5.1 Taking samples　Cut the fresh or dry samples into moderate size.

5.2 Fixation or soften　Soak samples into fixative for 12-24h. F.A.A fixative is commonly used. Dry samples could be soaked with water or glycerin without fixation.

5.3 Rinsing　Wash samples several times with 50% alcohol after fixation.

5.4 Dehydration　The fixed sample is soaked in different concentrations of ethanol and dehydrated. Ethanol concentration gradients are usually 50%, 70%, 80%, 90%, and 100% in that order. The immersion time of each gradient is 1-3h, depending on different materials.

5.5 Transparentizing　The sample is soaked in 50% xylene alcohol solution for 2-3h, then soaked in pure xylene twice for 1-2h each time.

5.6 Infiltration with paraffin　Generally, transparent samples are soaked in the mixture of paraffin and xylene at room temperature for 1-2 days. The solid pure paraffin was placed in a incubator to melt, and the temperature of the incubator was generally set to 2-3℃ higher than the melting point of paraffin. The sample is transferred from the mixture of paraffin and xylene to the melted paraffin. Replace pure paraffin every 2h for 3 times. The immersion time depends on the characteristics and size of the sample.

5.7 Embedding　Pour the samples into the paper box together with paraffin while hot. Adjust the position of the samples with heated tweezers，and then cool down to solidify (can be put in cold water).

5.8 Slicing   Take the embedded wax blocks and shape them into cuboids with parallel upper and lower sides. The block is fixed to the rotary microtome and cut into 5-10μm thick continuous paraffin sections. Remove the paraffin sections carefully with a brush.

5.9 Sticking   Smear a little gelatin adhesive onto a clean slide, then add drops of water. Transfer the paraffin sections to water with tweezers. The commonly used adhesive is Haupt paste or egg whites. Preservatives such as sodium salicylate can be added to the adhesive to improve the storage time.

5.10 Baking or developing   The slides were placed on the heated plate at 40℃, and the paraffin sections above the slides were heated and stretched out. Then transfer the slides to a 37℃ incubator to dry for more than 12h.

5.11 Dewaxing   Soak slides with paraffin sections in pure xylene for about 10 minutes to completely dissolve the paraffin for staining.

5.12 Staining   Slides are soaked in different concentrations solution. The solution are usually 50% xylene alcohol, 95%、85%、70%、50%、30% ethanol and pure water in turns. Soaking time of each solution is about 5-10min. Soak slides in 1% safranine solution for 1-24h. After taking out and washing it, soak the slides in 30%、50%、70% ethanol by turns to dehydration. Soak slides in 0.5% fast green solution for 10-40 sections.

5.13 Sealing   The stained slices were dehydrated with gradient ethanol and transparent with xylene. The last step is that sealant was added on slide with sample, and then a clean cover glass was put down to seal the slices. Sealant commonly used is Canada gum or neutral gum.

6. Biology drawing

Drawing plays an important role of the experimental report of medicinal botany. In the process of drawing, it is necessary to keep the objectivity and accuracy of the description of objectives. When drawing, it should be clear in primary and secondary, unified in picture, lively, supplemented by certain artistry. It is necessary to avoid smudging, and use the density of dots and shading lines to represent the light and shade contrast and quality difference of plant organs and tissues, when drawing. All drawing notes shall be marked with horizontal straight lines on the right side of the drawing. When there are many contents marked, broken lines can be used, but they must be in order and consistent. All drawing notes shall be kept in one column. Arc, arrow line and cross line are not available for annotation. The title of the drawing and the name and position of the materials used should be written below the drawing. When drawing the microstructure diagram, the magnification shall be indicated.

6.1 Drawing tools   Plant scientific drawing includes morphological drawing and structural drawing. Generally, the drawing is ink drawing with simple tools. Pen, the ink drawing of the experimental report is generally drawn with pencil. Pencils of HB (the outline is clear and easy to modify）, 2H, 3H (easily to describe the morphological characteristics of plants and medicinal materials) are commonly used. Paper, blank paper without lines is commonly used for the experimental report. Ruler, caliper, scale, drawing rubber etc. are needed too.

6.2 Drawing method of medicinal plants and medicinal materials   The drawing of medicinal plants and medicinal materials should be presented in the form of ink drawing in combination with the characteristics of specimens.

When drawing plants, the main body of plants usually should be drawing first, reserve the painting space of each organ, and make a detailed description according to the characteristics of each important

organ. For some important features, the magnification can be drawn to show their features.

6.3 Drawing method of microstructure drawing   Microstructural drawing is the structural drawing of medicinal plants. The microstructural features of cells, tissues and organs (transverse, longitudinal and surface) of plants are observed and mapped by microscopes, including and detail drawing.

6.3.1 Tissue diagram   Tissue diagram drawing is a biological drawing method which uses various simple lines to represent the boundaries of various tissues and various specific symbols to represent tissue types and characteristics (Figure 2, see them in Chinese part). In the process of drawing, it is not necessary to draw the shape of plant cells.

6.3.2 Detail drawing   Detail drawing is a structural drawing of cells, tissues and organs drawn based on the detailed shape, characteristics and distribution characteristics of cells. Detail drawing includes organ structure drawing, dissociated tissue structure drawing, tissue powder structure drawing, etc. It is required to outline the internal characteristics of organs, tissues and cells in detail, such as thickness of cell wall, thickened lamina, type of conduit, type of mesophyll cell, type of post inclusion, etc. When drawing organ structure, the whole or partial organ map can be drawn according to requirements.

6.3.3 Drawing method   Place the drawing required for drawing on the right side of the microscope, observe the image in the microscope with the left eye, and complete the drawing with the right eye. When drawing, you need to select the parts with obvious features. First, use 2H pencil to sketch on the drawing paper, then carefully observe the interior of the specimen, and use HB pencil to draw the drawing.

7. Methods of microchemical identification

7.1 Definition of microchemical identification   Microchemical identification is one of important methods used to display certain chemical components (starch grains, proteins, fats, etc.) in organs, tissues or cells in slices by using chemical reactions or staining methods of plant materials made by freehand sectioning. It is a rapid method to characterize and locate the shape, color, and distribution of these substances directly under a microscope.

7.2 Application scope of microchemical identification   This method is not only widely used in study of plant organs, tissues and ergastic substances, but also used to distinguish cell structures.

7.3 Precautions for microchemical identification

7.3.1 Use fresh and healthy plant materials.

7.3.2 Plant tissue section should be prepared as soon as possible after the material is taken. Frozen sections or freehand sections are most suitable for making thin slices.

7.3.3 The slices should not be too thin. If the content of the chemicals is too low in the thin slices, the chemical reaction will not be obvious. The thickness of material is 20-40μm.

7.3.4 For a substance tested, the selected reagent and chemical reaction method must be specificity.

7.3.5 The number of samples, or the number of observations and identifications, must be repeated, and experimental errors should be ruled out as much as possible.

7.3.6 For microchemical identification, a control test should be set up.

7.3.7 The reaction-treated section and the control section should be as close as possible, It is better to use the former slices for the reaction treatment section and the latter slices for the control section.

7.3.8 Record and take pictures of the observations at any time.

7.4 Common microchemical identification methods　Here we mainly introduce the method of microchemical identification of the chemical components of the cell wall and the ergastic substance in cells.

7.4.1 Microchemical identification of the ergastic substance

7.4.1.1 Identification of starch　Starch is the most important storage substances in plants. It usually exists in plant cell in the form of starch grains, which show various characteristics in different plant cells. The commonly reagent for identifying starch is Iodine-Iodination kalium. Starch reacts with iodine to form iodinated starch, which exhibits a special blue or brown-red reaction. It should be noted in the experiment that the iodine solution should not be too concentrated.

7.4.1.2 Identification of inulin　Inulin is formed by the aggregation of fructose molecules, and is mainly present in the cells of Cornaceae, Asteraceae, Gentianaceae and Campanulaceae families. Inulin is easily soluble in water and insoluble in ethanol. In order to observe inulin crystals, the experimental materials should be immersed in the 95% ethanol solution one week in advance to precipatate the inulin crystals. Then add a 10% solution of α-naphthol in ethanol to the material containing inulin, and then add a drop of 80% sulfuric acid. It could be observered that inulin appears purple-red under the microscope, but will soon dissolve.

7.4.1.3 Identification of protein (aleurone grain)　The protein stored in plant cells often exists in the form of solid particles—aleurone grain. The most common reagent to identify protein is Iodine-Iodination kalium. The method is simple and can be performed simultaneously to identify the starch, the protein is dyed yellow, but the starch grains is dyed bluish-violet.

7.4.1.4 Identification of fat and fat oil　The Sudan Ⅲ or alcohol solution of Sudan Ⅳ are often used to identify fat and fat oil, and fat or fat oil droplets are stained orange-red. However, this method cannot be used as evidence for the presence or absence of fat,, because resin, volatile oil and keratinized cell walls can also be stained red by Sudan Ⅲ or Sudan Ⅳ alcohol solutions, but the fat is more obvious after staining.

7.4.1.5 Identification of crystals　The crystals in plant cells are calcium oxalate and calcium carbonate crystals, among which the calcium oxalate crystals are the most common the morphological characteristics and the locations of which can be used as one of the basis for identification of Chinese herbal medicines. The two can be identified by acetic acid, dilute hydrochloric acid, or concentrated sulfuric acid. Calcium oxalate crystal does not dissolve in dilute acetic acid, but dissolves in dilute hydrochloric acid or concentrated sulfuric acid without bubbles, while calcium carbonate crystals dissolves in acetic acid and dilute hydrochloric acid or concentrated sulfuric acid and CO_2 bubbles will generate.

7.4.2 Microchemical identification of cell wall chemical components

7.4.2.1 Identification of cellulose　The cell wall is mainly composed of cellulose, and the commonly used identification method is Zinc chloride-Iodine method. The materials are sliced by freehand section, and the thinner is placed on a glass slide. 1-2 drops of Zinc chloride-Iodine can be added for staining, and the cellulose was stained blue or purple under a microscope. Iodine-Iodination kalium and 66% concentrated sulfuric acid also can be used for staining.

7.4.2.2 Identification of lignified cell walls　Lignification means that the cell wall is filled with lignin, which can increase the hardness of the cell wall and increase the mechanical force of the cell population. The materials can be cut into thin slices and stained by 1-2 drops of concentrated hydrochloric acid and phloroglucinol solution. Under the microscope, it can be observed that the lignified cell wall is purple or cherry red after staining.

7.4.2.3 Identification of suberized cell walls   Suberization means the cell wall is filled with the fatty compound—suberin, which has a protective effect on the internal tissues of the plant. Suberized cell walls are often yellow-brown and are dead cells. The Sudan III test solution is commonly used for staining. Under the microscope, the corked walls can be observed to be orange-red or red.

7.4.2.4 Identification of cutinization   Fatty compounds produced by protoplasts in cells is colorless and transparent, which will accumulate inside and on the surface of the cell wall, which will make the cell wall cutinization and form the cuticle, both of them can prevent the evaporation of water and the invasion of microorganisms to protect the internal tissues of plants. The identification of cutinization is the same as that of suberization, adding Sudan Ⅲ into the material to show tangerine or red.

7.4.2.5 Identification of mucoid cell walls   Mucoidification is a change in which the components such as pectin and cellulose contained in the cell wall and become mucus. The epidermal cells of the seed of *Plantago asiatica* L.,*Brassica juncea* (L.) Czern. et Coss., *Linum usitatissimum* L. and the fruit of *Salvia miltiorrhiza* Bge. all have mucusified cells. It is often identified by using sodium rosolate ethanol solution, and the cell wall is dyed rose-red under the microscope.

7.4.2.6 Identification of mineralized cell walls   The mineralization is due to the addition of siliceous (silicon dioxide or silicate) or calcium in the cell wall, which can enhance the mechanical support ability of plants. There are a lot of silicate in the stems and leaves of Gramineae and the stems of Equisetum hyemale L. and the mineralized cell walls of diatom. Generally, sulfuric acid or acetic acid is added for identification. Silicified cell walls are insoluble in both acid. This method can be used to identify the presence of siliceous.

三、药用植物标本的制作方法

（一）腊叶标本的制作

腊叶标本，就是把采集到的新鲜植物全株或一部分经压制干燥、消毒、上台等工序制成的干燥植株标本，也叫"压制标本"。此法对种子植物、蕨类及苔藓类植物均适用，其制作过程大致如下。

1. 器材与材料　剪刀、密封袋、吸水纸、标本夹（或夹板）、粗绳、台纸、标签纸等。

2. 压制前处理　宜选用形态完好、大小适宜，且具有繁殖器官的新鲜植株，将多余的枝叶进行修剪，去掉泥土等杂物。细小的种子及花粉宜单独装入小袋中，与该植物标本放在一起。粗大的地下部分宜切割一小部分，或与其地上部分的标本编成同一号码后另行干燥。特别注意，有鳞茎、根茎或块根等的植物，应该连同他们地上的部分挖取后成标本。

3. 做好采集记录　采集材料的时候，需要按标本采集记录所列各项认真填写，详细记录采集地点，植物生境，所处的经纬度、海拔以及植物相关形态特征，特别是压制后容易改变的特征，如花、果实的颜色、叶片上的白粉、容易脱落的毛茸等要详细记录。

4. 挂号牌　写明采集时间、地点以及采集人姓名。雌雄异株植物需要分开编号，写明是同一种的雌株和雄株。号牌上所列各项填写后，应拴在植物标本上。号牌上的采集号必须与采集记录的标本号一致。

5. 压制腊叶标本　压制腊叶标本时，首先将标本架或标本厚夹板的一片放平，先在上面放一块与标本夹大小相当的瓦楞纸，然后在上面放吸水纸，吸水纸的多少视采集材料的含水量而定。将经过挑选并拴好号牌的新鲜植物摆在吸水纸上面，其上面再放几张吸水纸，如此反复操作。最后将另一面标本夹（或夹板）盖上，用绳子捆紧，放在通风良好处干燥。期间需要反复更换吸水纸，并对半干的标本整形，直到标本完全干燥为止。

6. 标本的消毒　野外直接采集的标本往往带有害虫卵或霉菌孢子，故在标本上台前必须进行消毒，消毒方法如下：配制的2‰~5‰的氯化汞（$HgCl_2$）乙醇溶液适量，将溶液倒入搪瓷盘内，将干燥的标本浸入氯化汞乙醇溶液中5~10min，然后用竹夹取出，放在干的吸水纸中，当乙醇挥发后，$HgCl_2$会留在标本上，起到防虫防腐的作用。氯化汞有剧毒，操作时应注意房间通风，切忌用手直接操作，要带胶皮手套和门罩，操作后要洗手以免中毒。剩余消毒液要妥善保管。目前也有用冷冻方法进行标本的消毒。标本干后即可上台纸。

7. 标本的上台　已经消毒的标本用毛笔将胶水均匀刷在标本背面，花的部分不必上胶，以便解剖观察花的各部形态，然后移贴在台纸上（台纸的规格为39cm×27cm），可用约40cm×30cm的厚卡片纸稍加压力。放置待干时，应注意在左上角和右下角分别留出贴采集记录和鉴定签的位置，并于标本右上角盖印"$HgCl_2$消毒"字样，然后用线将叶片和植物粗壮部分穿钉固定在台纸上即成。

8. 标本的鉴定与存放　标本装订完毕后，需要进行准确鉴定。可参考《中国高等植物分科检索表》进行检索分科，确定科别后，再利用《中国植物志》《中国高等植物图鉴》以及其他地方植物志进行分属和分种鉴定。鉴定标本时，要利用放大镜和解剖镜对标本仔细观察，按照检索表的各条款项逐一核对，尤其注意花、果实、种子等繁殖器官的核对。

标本经过鉴定后，可将鉴定签贴在台纸的右下角，记录贴在左上角，最后在标本上面可加贴一张硫酸纸，以免标本互相摩擦损坏，这样即成为完整的标本。随后将同种植物标本用种夹放在一起，再按科属顺序放入标本柜中密闭保存。柜中可放入一些樟脑丸防虫，整个标本室可用硫酰氟或溴代甲烷熏蒸消毒，但消毒后要打开窗户通风数日方可进入。

（二）浸制标本的制作

浸制标本是通过将新鲜植物全株或器官浸泡在具有杀菌防腐作用的溶液里，达到较长时期保存为目的标本。制作时，将完整的植物或经过挑选和修剪的新鲜植物标本放进玻璃容器（通常用大型磨口玻璃标本瓶或磨口缸）中，然后加 10% 甲醛溶液或加 30%~50% 的乙醇和少许甘油，使之淹没植物标本，最后加盖封闭保存。在玻璃容器外面贴上标签。

浸制标本适用于新鲜花朵、各种地下茎、大型肉质果实及多汁的植物的长期保存。对于某些藻类、真菌等，应将每种植物分别浸制并装入不同的容器内，以免孢子相互混淆。浸制标本使用的固定液有 F.A.A. 固定液，5%~6% 的甲醛溶液，60%~70% 乙醇溶液，3%~5% 冰醋酸溶液等。

习题

Ⅲ　Preparation Method of Medicinal Plant Specimen

1. Preparing herbarium specimen

Herbarium specimen are dried plant specimens made from the whole or part of the collected fresh plants through pressing, drying, disinfection, and stage processing. They are also called "dehydrated specim". This method is applicable to seed plants, ferns and bryophytes, and its preparation method process is as follows.

1.1 Equipment and materials　Scissors, sealing bags, water-absorbing paper, specimen holder (or splint), thick rope, label paper etc. .

1.2 Pre-press treatment　Fresh plants with intact form, suitable size, and reproductive organs should be selected. Excess branches and leaves are trimmed to remove dirt and other debris. Tiny seeds and pollen should be packed in small bags separately and placed together with the plant specimen. The coarse underground part should be cut into a small part, or the same number as the above-ground specimen should be cut and dried separately. In particular, plants with bulls, rhizomes, or tubers, etc., should be excavated together with the ground part.

1.3 Make collection records　Take a careful record according to the items given by the outdoor directions, such as a clear collection site, plant habitat, latitude, longitude, altitude, and plant-related morphological characteristics, especially those that are easily changed after preparation, such as the color of the flower and fruit, white powder on the leaves, and hairy that easily fall off should be recorded in detail.

1.4 Specimen tag　State the time and place of collection and the name of the collector. Dioecious plants need to be numbered separately, indicating that they are the same female and male plants. After filling in the items listed on the collection records, the tag should be tied to the plant specimen. The collection number on the tag must be the same as the specimen number on the collection record.

1.5 Pressed fresh plant samples　Firstly lay down the sample shelf or plywood and put several pieces of water-absorbing paper on it; secondly put another piece of paper on the water-absorbing paper to sandwich your sample which has already been chosen and numbered together with the collection records; then, you need to put some water-absorbing paper on it again. Repeat this process again. Finally, cover the whole accumulation with another shelf or plywood and bundle them tightly. Place it in a well-ventilated place to dry. During this period, it is necessary to repeatedly replace the absorbent paper and shape the semi-dry specimen until the specimen is completely dry.

1.6 Disinfection of specimens　Specimens collected directly in the field often contain harmful insect eggs or mold spores. Therefore, the specimens must be disinfected. The disinfection method is as follows: firstly, the appropriate amount of 2‰-5‰ of mercury ($HgCl_2$) ethanol solution was prepared, then poured into the enamel tray, and the dried specimen was immersed in above ethanol solution for 5-10min. Then the specimen was taken out with a bamboo clip and placed on dry blotting paper. When the ethanol volatilized, $HgCl_2$ would remain on the sample to prevent insects and prevent corrosion. Because mercury is highly toxic, operation should pay attention to ventilation in the room, avoid direct operation by hand, to take rubber gloves and door, after operation to wash hands to avoid poisoning. Keep the remaining disinfectant in a safe place. Freezing treatment is also currently used to sterilize specimens. After the specimen is dried, it can be sticked on the hard paper.

1.7 Stick the specimen on hard paper   Brush the glue on the back of the specimen with a brush. The flower part does not need to be glued, in order to dissect and observe the shape of each part of the flower. Then transfer the specimen to the hard paper (the size of the paper is 39cm × 27cm), and you can apply a little pressure with a thick card paper of about 40cm × 30cm. When it is placed to be dried, it should be noted that the upper left corner and the lower right corner should be set apart to paste the collection record and identification label, and the words "$HgCl_2$ disinfection" should be stamped on the upper right corner of the specimen, and then the leaf and the thick part of the plant should be fixed on the table paper with thread.

1.8 Specimen identification and storage   After the specimens are sticked on the hard paper, accurate identification is required. You can refer to the *Higher Plant Family Index of China* to search the family of the specimen. After determining the family, you can use the *Flora of China*, *Higher Plant Picture of China* and other local flora for genus and species indentification. When identifying specimens, carefully observe the specimens with a magnifying glass and a dissecting mirror, and check them one by one according to the terms of the search form. Pay particular attention to the reproductive organs such as flowers, fruits and seeds.

After the specimen has been identified, the identification tag can be affixed to the lower right corner of the paper, and then the collection record on the upper left corner. Finally, a piece of sulfuric acid paper can be affixed to the specimen to prevent the specimens from being rubbed and damaged. Subsequently, then the specimens of the same species should be put together in a species folder, and then placed in the specimen cabinet in a sealed order in order of families and genera. Some camphor pills could be put in the cabinet to prevent insects. The entire specimen room can be fumigated with sulfuryl fluoride or bromomethane, but after disinfection should open a window to ventilate for a few days before entering.

2. Preparation of immersion specimens

The immersion specimen is preserved for a long period of time by immersing the whole plant or organ of fresh plants in a solution with bactericidal and antiseptic effects. When making, put the whole plant or fresh plant samples that have been selected and trimmed into glass containers (usually large ground glass bottles or mortar jars), and then add 10% formaldehyde solution or 30%-50% alcohol and a little glycerin to make it submerged plant specimens, finally sealed preservation. Label the outside of the glass container.

The immersion specimen is suitable for long-term preservation of fresh flowers, various underground stems, large fleshy fruits and juicy plants. For some algae, fungi, etc., each plant should be soaked separately and put into a different container, so as not to confuse the spores. The fixatives used for immersion specimens also include F.A.A. fixatives, 5%-6% formaldehyde aqueous solution, 60%-70% ethanol aqueous solution, 3%-5% glacial acetic acid aqueous solution, etc.

实验部分

实验一　根的形态、类型和变态

目的要求

1. **掌握**　识别根的外部形态和类型。
2. **熟悉**　常见药用植物变态根的形态、特征和类型。

实验材料

桔梗、蒲公英、大豆、薏苡、葱、徐长卿、芍药、菘蓝、白芷、何首乌、天门冬、萱草、龟背竹、吊兰等新鲜植物的根或腊叶标本，菟丝子、桑寄生或槲寄生带寄主的标本，络石标本。

仪器用品

镊子、放大镜、解剖镜。

内容与方法

（一）根的形态特征和类型

1. **直根系**　观察桔梗、蒲公英、大豆的根系，注意分辨主根、侧根和纤维根。
2. **须根系**　观察薏苡、葱、徐长卿的根系，注意观察根系是如何形成的，是否有主根、有无不定根。

（二）根的变态

1. **贮藏根**　观察桔梗、芍药、菘蓝、白芷等植物的根，可见其直根膨大成不同形状，这种变态根主要适应于贮藏大量的营养物质。
2. **块根**　观察何首乌、天门冬、萱草等植物的根，注意与贮藏根的区别。
3. **气生根**　观察龟背竹、吊兰露在空气中的不定根；观察薏苡伸入土壤中的不定根。
4. **寄生根**　观察带有寄主的菟丝子、桑寄生或槲寄生标本，它们的根均伸入了寄主的茎内。其中菟丝子不含叶绿体，不能制造养料。
5. **攀援根**　观察络石，注意由茎上产生能攀附他物的不定根。

💬 **实验报告**

把观察到的变态根类型列表记录下来。

📝 **思考题**

1. 定根和不定根有什么区别？
2. 植物的根为什么会发生变态？这种变态有何意义？

习题

Experimental Section

Experiment 1   Morphology, Type and Modifications of Root

Aim and demand

1. Grasp and recognize the external morphological characters and types of root.
2. Know the morphological characters and types of modified root of common medicinal plants.

Experiment materials

The materials of root: Fresh roots or dehydrated specimen of Jiegeng (*Platycodon grandiflorum* (Jacq.) A. DC.), Pugongying (*Taraxacum mongolicum* Hand. –Mazz.), Dadou (*Glycine max* (L.) Merr.), Yiyi (*Coix lacryma-jobi* L. var. ma-yuen (Roman.) Stapf), Cong (*Allium fistulosum* L.), Xuchangqing (*Cynanchum paniculatum* (Bunge) Kitag.), Shaoyao (*Paeonia lactiflora* Pall.), Songlan (*Isatis indigotica* Fort.), Baizhi (*Angelica dahurica* (Fish. Ex Hoffm.) Benth et Hook. f.), Heshouwu (*Polygonum multiflorum* Thunb.), Tianmendong (*Asparagus cochinchinensis* (Lour.) Merr.), Xuancao (*Hemerocallis fulva* (L.) L.), Guibeizhu (*Monstera deliciosa* Liebm.), Diaolan (*Chlorophytum capense* (L.) Voss.), samples with host of a parasite of Tusizi (*Cuscuta chinenses* Lam.), Sangjisheng (*Taxillus chinensis* (DC.) Danser) or Hujisheng (*Viscum coloratum* (Kom.) Nakai), sample of Luoshi (*Trachelospermum jasminoides* (Lindl.) Lem.).

Instrument and appliances

Tweezers, magnifying glass, anatomical lens.

Contents and Procedures

1. The morphological characters and types of root

1.1 Taproot system   Observe the roots of *Platycodon grandiflorum*, *Taraxacum mongolicum* and *Glycine max*, to distinguish main root, lateral root and fibrous root.

1.2 Fibrous root system   Observe the root system of *Coix lacryma-jobi*, *Allium fistulosum* and *Cynanchum paniculatum*, and paying attention that how these roots form the root system, and if it contains main root or adventitious root.

2. Root modifications

2.1 Storage root　Observe the roots of *Platycodon grandiflorum*, *Paeonia lactiflora*, *Isatis indigotica* and *Angelica dahurica*. The main roots of these plants expand into different forms which are mainly suitable for storing a large number of nutrients.

2.2 Root tuber　Observe the root tubers of *Polygonum multiflorum*, *Asparagus cochinchinensis* and *Hemerocallis fulva*, and comparing root tuber with storage root.

2.3 Aerial root　Observe adventitious roots that exposed in the air of *Monstera deliciosa* and *Chlorophytum capense*, and roots of *Coix lacryma-jobi* that extend into the earth.

2.4 Parasitic root　Observe sample with host of a parasite of *Cuscuta chinenses*, stem of *Taxillus chinensis*, *Viscum coloratum*, but *Cuscuta chinenses* doesn't have chloroplast to produce nutriment.

2.5 Climbing root　Observe the climbing adventitious root of *Trachelospermum jasminoides*. Note the origin of adventitious roots.

Experiment Report

List the type of root modifications observed.

Questions

1. What are the differences between the normal root and adventitious root?
2. Why does plant produce root modification? What's the significance of root modification?

实验二　茎的形态、类型和变态

目的要求

1. **掌握**　识别茎的外部形态和类型。
2. **熟悉**　常见药用植物变态茎的形态、特征和类型。

实验材料

葡萄、薄荷、益母草、天竺葵、地锦、常春藤、姜、黄精、马铃薯、半夏、荸荠、慈姑、洋葱、百合、天门冬、皂荚、栝楼等新鲜植物的茎或腊叶标本。

仪器用品

镊子、放大镜、解剖镜。

内容与方法

（一）茎的形态特征和类型
1. **茎的外形**　观察葡萄、薄荷、益母草的茎，观察并区别茎与根的形态差异。
2. **茎的类型**　观察天竺葵、地锦、葡萄、常春藤的茎，注意区分不同类型的攀援茎。
（二）茎的变态
1. **地下变态茎**
（1）根状茎　观察姜、黄精等植物的根状茎，注意观察节和节间、芽和鳞片状退化的叶。
（2）块茎　观察马铃薯、半夏等植物的块茎。注意观察块茎上的芽。
（3）球茎　观察荸荠、慈姑等植物的球茎。
（4）鳞茎　观察洋葱、百合等植物的鳞茎。注意观察鳞茎盘、鳞叶和不定根。
2. **地上变态茎**
（1）叶状茎　观察天门冬等植物的叶状茎。
（2）刺状茎　观察皂荚等植物的刺状茎。
（3）茎卷须　观察葡萄、栝楼等植物的茎卷须。注意观察卷须发生的位置。

实验报告

把观察到的变态茎类型列表记录下来。

🖍 思考题

1. 植物的根和根状茎有什么区别？
2. 举例说明植物茎的形态特征与生理功能之间的关系。

Experiment 2   Morphology, Type and Modification of Stem

Aim and demand

1. Grasp the morphological characters and types of stem.
2. Grasp the morphological characters and types of modified stem of common medicinal plants.

Experiment materials

The materials of stem: Fresh stems or dehydrated specimen of Putao (*Vitis vinifera* L.), Bohe (*Mentha haplocalyx* Briq.), Yimucao (*Leonurus japonicus* Houtt.), Tianzhukui (*Pelargonium hortorum* Bailey), Dijin (*Parthenocissus tricuspidata* (Sieb. et Zucc.) Planch.), Changchunteng (*Hedera nepalensis* K. Koch var. *sinensis* (Tobl.) Rehd.), Jiang (*Zingiber officinale* Roscoe), Huangjing （*Polygonatum sibiricum* Delar. *ex* Redoute), Malingshu (*Solanum tuberosum* L.), Banxia (*Pinellia ternata (Thunb.)*Breit.), Biqi (*Eleocharis dulcis* (N. L. Burman) Trinius ex Henschel), Cigu (*Sagittaria trifolia* L. var. *sinensis* (Sims.) Makino), Yangcong (*Allium cepa* L.), Baihe (*Lilium brownii* F. E. Brown ex Miellez var. *viridulum* Baker), Tianmendong (*Asparagus cochinchinensis* (Lour.) Merr.), Zaojia (*Gleditsia sinensis* Lam.), Gualou (*Trichosanthes kirilowii* Maxim.).

Instrument and appliances

Tweezers, magnifying glass, anatomical lens.

Contents and Procedures

1. The morphological characters and types of stem

1.1 Shape of stem   Observe the stems of *Vitis vinifera*, *Mentha haplocalyx* and *Leonurus japonicus*, to distinguish the differences between stem and root.

1.2 Type of stem   Observe the stem of *Pelargonium hortorum*, *Parthenocissus tricuspidata*, *Vitis vinifera*, *Hedera nepalensis*, and pay attention to difference types of climbing stem.

2. Stem modifications

2.1 Modifications of underground stem

2.1.1 Rhizome   Observe the rhizomes of *Zingiber officinale*, *Polygonatum sibiricum*, to find out the nodes, internodes, buds and degenerative scale leaves.

2.1.2 Tuber   Observe the tubers of *Solanum tuberosum*, *Pinellia ternata*, and pay attention to their buds.

2.1.3 Corn   Observe the corns of *Eleocharis dulcis, Sagittaria trifolia* var. *Sinensis*.

2.1.4 Bulb   Observe the bulbs of *Allium cepa, Lilium brownii* var. Viridulum and pay attention to their thickened bulb, scaly leaves and adventitious roots.

2.2 Modifications of overground stem

2.2.1 Cladophyll   Observe the leafy stem of *Asparagus cochinchinensis*.

2.2.2 Spiny stem   Observe the spiny stem of *Gleditsia sinensis*.

2.2.3 Stem tendril   Observe the stem tendril of *Vitis vinifera, Trichosanthes kirilowii,* to find out the location of tendril.

Experiment report

List the type of stem modifications.

Questions

1. What are the differences between root and rhizome?

2. Give examples of the relationship between the morphological characteristics and physiological functions of plant stems.

实验三 叶的形态、类型和变态

目的要求

1. **掌握** 叶的形态、类型及变态叶的形态和类型。
2. **掌握** 叶的组成，叶片的形状及叶端、叶基、叶缘；单叶与复叶的区别；复叶的类型；叶序。
3. **识别** 常见的叶类药用植物。

实验材料

1. **鲜活植物** 女贞、贴梗海棠、枇杷、艾、扶桑、木槿、紫草、忍冬、刺黑珠、黄精、银杏、薄荷的叶；蒲公英、马蹄莲、龟背竹的苞叶；洋葱、百合、麻黄的鳞叶；仙人掌、构骨、刺槐的叶刺与托叶刺；豌豆及菝葜的卷须。
2. **腊叶标本** 月季、槐或紫藤、合欢、南天竹、酢浆草、白花草木樨、刺五加、甘草、葛、酸橙等植物的叶。

仪器用品

镊子、放大镜、解剖镜、解剖用具、显微镜。

内容与方法

（一）叶的形态

1. **叶的组成** 观察女贞、贴梗海棠、扶桑、木槿等植物叶的形态，分辨叶片、叶柄、托叶；注意其叶端、叶基、叶缘的形态和脉序的类型。

2. **单叶和复叶** 观察月季、槐或紫藤的羽状复叶，与扶桑、木槿叶比较。注意观察复叶的小叶柄没有腋芽，而总叶柄和单叶的叶柄均具有腋芽。

（二）叶的类型

1. **复叶的类型** 观察合欢、紫藤、南天竹、酢浆草、白花草木樨、刺五加、甘草、葛藤、酸橙、槐、刺五加、天南星、车轴草等植物的叶，判断其为何种复叶。

2. **叶序** 观察紫草、忍冬、刺黑珠、黄精、银杏、薄荷的茎枝，判断其叶序类型。

（三）叶的变态

1. **苞叶** 观察蒲公英、马蹄莲、龟背竹的花序，指出哪一部分是叶的变态？

2. **鳞叶** 观察洋葱、百合、麻黄的茎，找出变态的叶，区别肉质鳞叶和膜质鳞叶。

3. **叶刺** 观察仙人掌、三颗针、构骨、洋槐等植物，注意区别叶刺与托叶刺。

4. **卷须叶** 观察豌豆由复叶顶端二、三对小叶变成的卷须及菝葜由托叶变成的卷须。

实验报告

1. 绘一片完全的外形图，注明各个部分名称。
2. 记录观察到的变态叶及叶序类型。

思考题

1. 如何区分单叶和复叶？常见的复叶有哪些类型？
2. 叶的变态类型有哪些？

Experiment 3   Morphology, Type and Modification of Leaf

Aims and Requirements

1. Grasp the morphology and type of leaf and leaf modification.

2. Grasp the composition and shape of leaf, leaf apex, leaf base, leaf margin, the distinction between simple leaf and compound leaves, type of compound leaves, and phyllotaxy.

3. To identify and be familiar with the types of medicinal plants that use leaves as medicine.

Experiment materials

Living plant: leaves of Nvzhen (*Ligustrum lucidum* Ait.), Tiegenghaitang (*Chaenomeles speciosa (*Sweet) Nakai), Pipa (*Eriobotrya japonica* (Thunb.) Lindl.)*,* Ai (*Artemisia argyi* Levl. et Vant.), Fusang (*Hibiscus rosa -sinensis* L.) and Mujin (*Hibiscus syriacus* L.), Zicao (*Lithospermum erythrorhizon* Sieb. et Zucc.), Rendong (*lonicera japonica* Thunb.), Ciheizhu (*Berberis julianae* Schneid.), Huangjing (*Polygonatum sibiricum* Delar. ex Red.). Yinxing *(Ginkgo biloba* L.), Bohe (*Mentha haplocalyx* Briq.); the bract of Pugongying (*Taraxacum mongolicum* Hand. -mazz.), Matilian (*Zantedeschia aethiopica* (L.) Spreng.), Guibeizhu (*Monstera deliciosa* Liebm.); the scale leaf of Yangcong (*Allium cepa* L.), baihe (*Lilium Brownii* F. E. Brown. var. *viridulum* Backer), Mahuang (*Ephedra sinica* Stapf）; the leaf thorn of Xianrenzhang (*Opuntia dillenii* (Ker -Gael.) Haw.), Gougu (*Hex cornuta* Lindl.), Cihuai (*Robinia pseudoacacia* L.); the leaf tendril of Wandou (*Pisum sairvum* L.) and Baqia (*Smilax sieboldii* Miq.).

Herbarium specimen: leaves of Yueji (*Rose chinensis* Jacq.), Huai (*Sophora japonica* L.), Ziteng (*Wisteria sinensis* (Sims.) Sweet.), Hehuan (*Albizzia julibrissin* Durazz.), Nantianzhu (*Nandina domestica* Thunb.), Cujiangcao (*Oxalis corniculata* L.), Baihuacaomuxi (*Melilotus albus* Desr.), Ciwujia (*Acanthopanax senticosus* (Rupr. et Maxim.)), Gancao (*Glyeyrrhiza uralensis* Fisch.), Ge (*Pueraria lobata* (Willd.) Ohwi), Suancheng (*Citrus aurantium* L.).

Instrument and appliance

Tweezers, magnifying glass, anatomical lens, dissector appliances, microscope.

Contents and Procedures

1. The external morphology of leaf

1.1 The composition of leaf   Observe the external morphology of leaves of Nvzhen (*Ligustrum lucidum* Ait.), Tiegenghaitang (*Chaenomeles speciosa* (Swee) Nakai), Fusang (*Hibiscus rosa -sinensis* L.), Mujin (*Hibiscus syriacus* L.). Distinguish blade, petiole and stipule. Pay attention to the characteristics of leaf apex, leaf base, leaf margin and type of venation.

1.2 Simple leaf and compound leaves Observe pinnately compound leaves of Yueji (*Rose chinensis* Jacq.), Huai (S*ophora japonica* L.) or Ziteng (*Wisteria sinensis* (Sims.) Sweet.). Compare Fusang (*Hibiscus rosa -sinensis* L.) with Mujin (*Hibiscus syriacus* L.). Note that the petiolule of compound leaves have no axillary buds, while the common petioles and single leaf have axillary buds.

2. The type of leaf

2.1 Type of compound leaves Observe the leaves of *Hehuan* (*Albizia julibrissin* Durazz.), Ziteng (*Wisteria sinensis* (Sims.) Sweet.), Nantianzhu (N*andina domestica* Thunb.), Cujiangcao (O*xalis corniculata* L.), Baihuacaomuxi (*Melilotus albus* Desr.), Ciwujia (*Acanthopanax senticosus* (Rupr. et Maxim.)）Harms. Gancao (*Glyeyrrhiza uralensis* Fisch.), Geteng (*Pueraria lobata* (Willd.) Ohwi), Suandieng (*Citrus aurantium* L.). Huai (*Sophora japonica* Linn.), Ciwujia (*Acanthopanax gracilistylus* W.W.Smith), Tiannanxing (*Arisaema heterophyllum* Blume), Chezhoucao (*Galium odoratum* (L.) Scop.). Determine the type of compound leaves.

2.2 Phyllotaxy Observe the stems of Zicao (*Lithospermum erythrorhizon* Sieb. et Zucc.), Rendong (*Lonicera japonica* Thunb.), Ciheizhu (*Berberis julianae* Schneid.), Huangjing (*Polygonatum sibiricum* Delar. ex Red.), Yinxing (*Ginkgo biloba* L.), Bohe (*Mentha haplocalyx* Briq.) and determine their type of phyllotaxy.

3. Leaf modifications

3.1 Bract Observe the inflorescence of Pugongying (*Taraxacum mongolicum* Hand. -mazz.), Matilian (*Zantedeschia aethiopica* (L.) Spreng.), Guibeizhu (*Monstera deliciosa* Liebm.) and point out where the leaf modification is.

3.2 Scale leaf Observe the inflorescence of Yangcong (*Allium cepa* L.), baihe (*Lilium Brownii* F. E. Brown. var. *viridulum* Backer), Mahuang (*Ephedra sinica* Stapf) and point out the abnormal leaf and distinguish fleshy scale leaf from membranous scale leaf.

3.3 Thorn leaf Observe Xianrenzhang (*Opuntia dillenii* (Ker-Gael.) Haw.), Sankezhen (*Berberis julianae* Schneid.), Gougu (*Hex cornuta* Lindl.), Yanghuai (S*ophora japonica* L.) and distinguish thorn leaf from thorn stipule.

3.4 Leaf tendril Observe the leaf tendril formed from two or three pairs of lobues at the top of compound leaves of Wandou (*Pisum sairvum* L.)*,* and the stipule turn into leaf tendril of Baqia (*Smilax sieboldii* Miq.).

💬 Experiment report

1. Draw the external morphology of a complete leaf, and note the name of every section.

2. List the abnormal leaves and the types of phyllotaxies observed.

✏️ Questions

1. How to distinguish the single leaf from compound leaves? How many types are compound leaves divided into?

2. How many types are leaf modifications divided into?

实验四 花的形态和花序

目的要求

1．**掌握** 花的解剖方法及使用花程式描述花的方法。
2．**熟悉** 花的组成。
3．**了解** 花序的概念并识别花序的类型。

实验材料

新鲜材料：蒲公英、荠菜、附地菜、玉兰、蓟、大戟等药用植物。

仪器用品

体视显微镜、镊子、解剖针、刀片、培养皿。

内容与方法

（一）花的组成

取一朵玉兰花，用解剖针和镊子由外向内、由下向上逐层剥离，按顺序将各部分放在白纸或培养皿内。

解剖时，应边解剖，边记录。

（1）花梗及花托：花梗长短及花托的形态。

（2）花萼：数量及离合情况。

（3）花冠：数量，离合情况，排列方式。

（4）雄蕊群：离合情况，数量，排列方式，雄蕊和花瓣的位置关系。

（5）雌蕊群：心皮数目、离合情况、子房位置、子房室数、胎座类型、胚珠数目。

另取 2~3 种不同类型的花，用上述方法观察，分析组成，并写出各自花程式。

（二）花序的类型

1．无限花序 观察荠菜的花序，花序轴较长，其上没有分枝；小花具近等长的花柄；开花顺序由下向上，而花序轴先端继续伸长并分化出新的花芽，这种花序为总状花序。

观察蒲公英、山楂、茴香、韭菜、知母、车前、无花果、蓟的花序，注意与总状花序有何异同，并指出其花序类型。

2．有限花序 观察附地菜的花序，注意其顶端先开放，开花顺序由上向下，因此花序轴不再继续伸长及产生花芽，只是在其下方同一侧产生侧轴，称为螺旋状聚伞花序。

观察大戟、石竹、丹参等植物的花序，注意与螺旋状聚伞花序的异同点，并指出其花序类型。

💬 **实验报告**

1. 任选一种花，绘制剖面图，注明各部分名称并写出花程式。
2. 绘制观察到的植物花序的简图。

思考题

1. 怎样判断子房位置？
2. 何为花序？如何区分有限花序和无限花序？

习题

Experiment 4　Floral Morphology and Inflorescence

Aim and demand

1. Grasp the anatomic method of flowers and describe a flower of angiosperms using a flower formula.
2. Be familiar with the compositions of flower.
3. Understand the concept of inflorescence and recognize the types of inflorescence.

Experiment materials

Fresh materials: Pugongying (*Taraxacum mongolicum* Hand.-Mazz.), Jicai (*Capsella bursa-pastoris* (L.) Medic.), Fudicai (*Trigonotis peduncularis* (Trev.) Benth. ex Baker et Moor), Yulan (*Yulania denudata* (Desrousseaux) D. L. Fu), Ji (*Cirsium japonicum* Fisch. ex DC.), Daji (*Euphorbia pekinensis*) Rupr.

Instruments and appliances

Stereo microscope, tweezers, dissecting needle, blade, culture dish.

Contents and Procedures

1. The composition of flower

Take a flower of *Yulania denudata*, and use an anatomic needle and tweezers to peel off it from the outside to the inside, from lower to upper, put every section on the newsprint or the culture dish.

Take notes, while dissecting a flower.

（1）Pedicel and receptacle: the length of the pedicel and the morphology of the receptacle.

（2）Calyx: the number of the sepal, separated or commissural.

（3）Corolla: the number of the petal, separated or commissural, and the arrangement of the petals.

（4）Androecium: the number of the stamen, separated or commissural, the arrangement of the stamens, and the relationship between the stamen and the petal in position.

（5）Gynoecium: the number of the carpel, separated or commissural, position of the ovary, the number of chambers in the ovary, the type of the ovary and the number of the ovule.

Moreover, observe 2-3 other types of flower by using the methods above and analyze the composition of the flower, and write the flower formula.

2. The type of inflorescence

2.1 Indefinite inflorescence　Observe the inflorescence of *Capsella bursa-pastoris*. The rachis of *Capsella bursa-pastoris* is long and unbranched; the pedicels of florets are about the same length;

the pedicels are subequal. The order of flowering is from downward to upward, while the apex of the inflorescence axis continues to elongate and differentiate into new flower buds. This inflorescence is raceme.

Observe the inflorescence of *Taraxacum mongolicum*, *Crataegus pinnatifida*, *Foeniculum vulgare*, *Allium tuberosum*, *Anemarrhena asphodeloides*, *Plantago asiatica*, *Ficus carica*, *Cirsium japonicum*. Note the similarities and differences with the raceme, and identify their inflorescence types.

2.2 Definite inflorescence Observe the inflorescence of *Trigonotis peduncularis*. The floret on the top of the rachis flowers first, and the order of flowering is from top to the bottom. So that the rachis no longer continues to elongate and produces buds, but in the same side of its lower side produces a lateral axis. This inflorescence is called spiral cyme.

Observe the inflorescence of *Euphorbia pekinensis*, *Dianthus chinensis*, *Salvia miltiorrhiza*, and compare them with the hericoid cyme, and identify their inflorescence type.

Experiment report

1. Draw the profile of any flower observed, note the name of each part and write the flower formula.
2. Draw the diagrams of the inflorescences of the flowers observed.

Questions

1. How to identify the position of the ovary.
2. What is inflorescence? How to distinguish the definite inflorescence and the indefinite inflorescence.

实验五　果实和种子的形态和类型

实验目的

掌握　果实的形态和类型；种子的形态和类型。

实验材料

1. **果实**　各种新鲜、干燥或浸制的果实标本。
2. **种子**　蓖麻、大豆等植物的种子。

实验仪器和用品

体式显微镜、放大镜、载玻片、盖玻片、镊子、解剖针、刀片、培养皿等。

实验内容

（一）果实形态和类型

1. **单果**　由单雌蕊或复雌蕊形成的果实，即1朵花只形成1个果实。

观察葡萄、桃、苹果、柠檬的果实，辨识其果皮的质地、雌蕊的心皮数、子房的室数和胎座的类型。

观察芍药、落花生、萝卜、曼陀罗、向日葵、玉米的果实，了解其果皮是否开裂、开裂方式及构成雌蕊的心皮数、胎座等特点。

2. **聚合果**　由1朵花中许多离生雌蕊形成的果实，每个雌蕊形成1个单果，聚生于同一花托上。

观察五味子、草莓、莲、八角、悬钩子的果实，识别其属于哪种聚合果类型，并注意离生雌蕊的数目、子房室数和胎座类型。

3. **聚花果**　由整个花序发育而成的果实，其中每朵花发育成1个小果，聚生在花序轴上，成熟后从花轴基部整体脱落。

观察菠萝、桑葚、无花果的果实，区分真果还是假果，注意其可食用部位的特点。

（二）种子形态结构和类型的观察

1. **有胚乳种子**　种子中有发达的胚乳，胚相对较小，子叶薄。

2. **无胚乳种子**　种子中胚乳的养料在胚发育过程中被胚乳吸收并贮藏于子叶中，故胚乳不存在或仅残留一薄层，这种种子通常具有发达的子叶。

在实验室内观察蓖麻或大豆种子，首先观察其外部形态，并注意各部位种阜、种孔、种脐、种脊和合点的位置及结构特点；然后用镊子轻轻去除种皮，并用刀片将种子进行纵切，观察二者的胚根、胚芽、胚轴和子叶的位置及结构特点。

💬 实验报告

1. 对上述观察到的果实进行列表分类，写出其主要特征。
2. 绘黄豆和蓖麻种子的外形及纵剖面图。

思考题

1. 如何区别菁葖果与荚果、角果和蒴果、瘦果与颖果？
2. 聚合果和聚花果在来源上有何不同，如何辨别？
3. 如何区别有胚乳种子和无胚乳种子？

习题

Experiment 5 | Morphology and Type of Fruit and Seed

Grasp the morphology and types of fruits and seeds.

1. **Fruits**   various of fresh, dried or immersed fruit specimens.
2. **Seeds**   the seeds of Bima (*Ricinus communis* L.) and Dadon (*Glycine max* (L.) Merr.).

Stereo microscope, magnifier, glass slide, cover slip, forceps, dissecting needle, blade, culture dish, etc.

1. The morphology and types of fruit

1.1 Single fruit   A plant fruit formed by a single pistil or a compound pistil, that is, one flower forms only one fruit.

Observe the fresh fruits, such as *Vitis vinifera* L., *Amygdalus persica* L., *Malus pumila* Mill., and *Citrus limon* (L.) Burm. f.. Identify the texture of its pericarp, the number of carpels, the number of ovary chambers and the type of placenta.

Observe the dried fruits such as *Paeonia lactiflora* Pall., *Arachis hypogaea* L., *Raphanus sativus* L., *Datura stramonium* L., *Helianthus annuus* L., *Zea mays* L. Investigate whether pericarp is dehiscent, how it dehisces, the number of carpel and the type of the placenta.

1.2 Aggregate fruit   a fruit formed by many isolated pistils, each pistil forming a single fruit, aggregating on the same receptacle.

Observe aggregate fruits such as *Schisandra chinensis* (Turcz.) Baill., *Fragaria ananassa* Duch., *Nelumbo nucifera* Gaertn., *Illicium verum* Hook. f., and *Rubus corchorifolius* L.f. Identify the type of aggregated fruit, the number of apocarpouspistil, the number of ovary chambers and type of placenta.

1.3 Polythalamic fruit   A fruit formed by the entire inflorescence, in which each flower develops into a small fruit, which is aggregated on the inflorescence axis and falls off from the base of the flower axis as a whole after maturation.

Observe polythalamic fruit such as *Ananas comosus* (L.) Merr., *Morus alba* L., *Ficus carica* L. Distinguish the true or false fruit, and notice the characteristics of its edible parts.

2. Observe the morphology and types of seeds

2.1 Albuminous seed   The seeds have a developed endosperm. The embryo is relatively small, and the cotyledons are thin.

2.2 Exalbuminous seed   The nutrients in the endosperm are absorbed by the endosperm during the development of the embryo and stored in the cotyledons, so there is no or only a thin layer remaining in the endosperm. Such seeds usually have developed cotyledons.

Observe the seeds of *Ricinus communis* L. or *Glycine max (L.)*Merr. in the laboratory. Notice their external morphology, the location and structural characteristics of caruncle, micropyle, hilum, raphe, and chalaza; then the seed coats were gently removed with tweezers and the seeds were cut longitudinally with a blade to observe the positions and structure of the radicle, germ, hypocotyl and cotyledon.

💬 Experimental Report

1. List the types of plant fruits above and write down their main characteristics.
2. Draw the appearance and profile of the seeds of *Ricinus communis* and *Glycine max*.

📝 Questions

1. How to distinguish follicle and legume, silique and capsules of silicle, achene and caryopsis?
2. What are the differences in the origin of aggregate fruit and polythalamic fruit, how to identify?
3. How to distinguish between albuminous seed and exalbuminous seed?

目的要求

1. **掌握**　藻类、菌类、地衣、苔藓和蕨类等孢子植物的主要特征。
2. **识别**　常见药用藻类、菌类、地衣、苔藓和蕨类植物。

实验材料

海带、甘紫菜、石莼、茯苓、冬虫夏草、灵芝、石耳、松萝、地钱、葫芦藓等植物标本；卷柏、海金沙等植物的腊叶标本；青霉菌、冬虫夏草子座、蕨叶、蕨原叶体等永久制片。

实验用品

解剖镜、放大镜、解剖器具等。

内容与方法

（一）藻类植物

观察各种藻类植物标本，总结藻类植物的主要特征。

1. **绿藻门**　石莼藻体为膜状体，由两层细胞组成，基部具有多细胞的固着器。全藻入药。

2. **红藻门**　甘紫菜藻体成薄膜状，遇水后，手摸有黏滑感，紫红色或淡紫红色。全藻入药。

3. **褐藻门**　海带植物体（孢子体）分三部分：成假根状的固着器、柄、带片。固着器分枝如根状，柄呈短茎状，柄以上是扁平叶状的带片，结构有表皮、皮层和髓之分。为中药"昆布"的基原植物之一。

（二）菌类植物

观察各种菌类植物标本，掌握菌类植物的主要特征。

1. **子囊菌亚门**　冬虫夏草下端虫体部分形成菌核，子座从其头部长出，呈棒状，上端稍膨大，具不育顶端。

观察子座横切片：子座周围长有子实体（子囊壳），子囊壳中生有多数子囊，各子囊中常有 2~8 个细长的子囊孢子，通常有 2 个成熟。

2. **担子菌亚门**

（1）**茯苓**　菌核呈不规则块状，表面灰褐色，粗糙有褶皱，断面白色。

（2）**灵芝**　子实体呈木栓质，菌盖半圆形或肾形，上面有光泽，红褐色，具横状环纹，下面白色，具小的管孔，菌柄生于侧面。

3. **半知菌亚门**　显微镜下观察青霉菌制片，分生孢子梗由多细胞组成，聚集呈扫帚状，末端分枝上生有成串的分生孢子。

（三）地衣植物

观察石耳、松萝的植物标本，掌握地衣植物的典型特征。

1. 石耳　叶状地衣。近圆形或稍不规则，革质。裂片边缘浅撕裂状，上表面褐色，近光滑，局部粗糙无光泽，或局部斑点状脱落而露生白色髓质。

2. 松萝　枝状地衣。地衣体长 20~40cm，表面灰绿色、草绿色。植物体丝状，二叉式分枝，基部分枝少，较粗，先端分枝多。

（四）苔藓植物门

观察苔藓的植物标本，总结苔藓植物的典型特征。

1. 地钱　植物体（配子体）呈扁平二叉分枝的叶状体，匍匐生长，生长点在二叉分枝的凹陷中，叶状体分为背腹两面，背面深绿色，腹面生有紫色鳞片和假根。雌雄异株。

2. 葫芦藓　植物体（配子体）无根，有茎和叶的分化。雌雄同株，异苞，雌枝产生颈卵器，雄枝产生精子器；孢子体分为孢蒴、蒴柄和基足三部分。

（五）蕨类植物门

1. 蕨叶　蕨叶由表皮、叶肉细胞和叶脉组成。叶背面着生大量孢子囊。显微镜下观察一个孢子囊结构，可见孢子囊具长柄，孢子囊壁由一层细胞组成。囊壁有一纵行内切向壁和侧壁增厚的细胞，称为环带。其中有少数不加厚的细胞称为唇细胞，唇细胞可使孢子囊开裂和散出孢子。

2. 蕨原叶体（配子体）　显微镜下原叶体为小而薄、绿色、略呈心形的叶状体，有背、腹面。腹面有假根，假根附近有精子器，在心形凹陷处有几个颈卵器。

3. 腊叶标本

（1）卷柏　主茎较长，根系密集成茎干状，小枝丛生在主茎顶端。孢子叶集生茎顶成孢子囊穗。

（2）海金沙　根状茎二叉状蔓生，被黑褐色毛。三至四回羽状复叶，叶轴几无毛。孢子细小，具瘤或疣状突起。

💬 实验报告

1. 写出藻类、菌类、地衣、苔藓和蕨类植物的主要特征。
2. 列表记录实验中所用的药用植物的名称、科名。

🖊 思考题

1. 蕨类植物的叶从来源和功能上可分为哪些类型？
2. 总结《中国药典（2020 年版）》收载的药用藻类、真菌和蕨类植物。

微课

微课

习题

Experiment 6　　Classification of Sporophytes

1. Grasp the main characteristics of algae, fungi, lichens, bryophyta and pteridophyta.
2. Identify common plants of algae, fungi, lichens, bryophyta and pteridophyta.

Experimental materials

Botanical specimen: Haidai (*Laminaria japonica* Aresch), Ganzicai (*Porphyra tenera* Kjellm.), Shichun (*Ulva lactuca* L.), Fuling (*Poria cocos* (Schw.) Wolf.), Dongchongxiacao (*Cordyceps sinensis* (Berk.) Sacc.), Ling (*Ganoderma lucidum* (Leyss ex Fr.) Karst.), Shier (*Umbilicaria esculenta* (Miyoshi) Minks), Songluo (*Usnea diffracta* Vain), Diqian (*Marchantia polymorpha* L.), Huluxian (*funaria hygrometrica* Hedw.) etc.

Dehydrated plant specimens: Juanbai (*Selaginella tamariscina* (P. Beauv.) Spring), Haijinsha (*Lygodium japonicum* (Thunb.) Sw.) etc.

Permanent slice: *Penicillium sp.*, Stroma of *Cordyceps sinensis* (Berk.) Sacc., The leaf of Jue (*Pteridium aquilinum* var. *latiusculum* (Desv.) Underw. ex Heller), prothallium of *Pteridium aquilinum* var. *latiusculum,* etc.

Experimental Appliances

Dissecting microscope, magnifying glass and dissect appliances, etc.

Contents and Procedures

1. Algae

Observe specimen of algae and summarize their main characters.

1.1 Chlorophyta　*Ulva lactuca* L. is flat membranaceous, and consists of 2 layers of cells. The base has a multicellular fixator. The whole plant is used as medicine.

1.2 Rhodophyta　*Porphyra tenera* Kjellm. is membranaceous. After meeting water, the hands feel sticky and slippery, purplish red or lavender red. The whole plant is used as medicine.

1.3 Phaeophyta　The plant of *Laminaria japonica* Aresch (sporophyte) is divided into three parts: fixator, stalk and strip. Branches of fixator are rootlike, stalks are short stemlike, and above stalks are flat leaflike bands with a structure of epidermis, cortex and pith. It is one of the original plants of traditional Chinese medicine 'kunbu'.

2. Fungi

Observe specimen of fungi and summarize their main characters.

2.1 Ascomycotina　The lower part of *Cordyceps sinensis* forms sclerotium, and the stroma grows out from the head of insect, which is rod-shaped. The upper part is slightly expanded and has a sterile top.

Observe the transverse slice of the stroma: there are sporophore (perithecium）around the stroma. There are many ascus in the perithecium, often 2-8 slender ascospores in each ascomycetes, usually two mature.

2.2 Basidiomycotina

2.2.1　Sclerotium of *Poria cocos* (Schw.) Wolf. is tuberous and irregular. The surface is taupe, rough and wrinkled. The section inside is white.

2.2.2　Sporophore of *Ganoderma luci*dum (Leyss ex Fr.) Karst. is corky. Pileus is semicircular or kidney-shaped, shiny above, reddish brown, with transverse annulus, white below, with small tubules, stipe arising laterally.

2.3 Deuteromycotina　Microscopic observation of *Penicillium* sp. The conidiophore is composed of many cells, clustered in a broomlike manner, with clusters of conidia on terminal branches.

3. Lichens

Observe specimen of *Umbilicaria esculenta* (Miyoshi) Minks and *Usnea diffracta* Vain. and summarize their main characters.

3.1 *Umbilicaria esculenta* (Miyoshi)Minks　Foliose lichen. Approximately circular or slightly irregular, leathery. Lobes margin is shallowly lacerated, upper surface brown, nearly smooth, locally rough, or locally flecked and exposed to white medulla.

3.2 *Usnea diffracta* Vain　Fruticose lichen. Lichen body length is 20-40cm, with the surface of gray-green, grass green. Plant body is filiform, binary branched. Base branch is less, thicker and apex branch is much.

4. Bryophyta

Observe specimen of bryophyta and summarize their main characters.

4.1 *Marchantia polymorpha* L.　The plant body (gametophyte) is a thallus with flat dichotomous branches, prostrate and growth points in the pits of the dichotomous branches. The thallus is divided into dorsal and ventral surfaces, abaxially dark green, with purple scales and pseudoroots on the ventral surface. Dioecious.

4.2 *Funaria hygrometrica* Hedw.　The plant body (gametophyte) is rootless and has stem and leaf differentiation. Monoecious. Female branch produces archegonium, and male branch produces antheridium. Sporophyte is divided into three parts: capsule, seta and foot.

5. Pteridophyta

5.1 The leaf of *Pteridium aquilinum* var. *Latiusculum*　It is composed of epidermis, mesophyll cells and veins. The abaxial surface of the leaf bears numerous sporangia. Microscopically, the structure of a sporangium shows that the sporangium has a long stalk and the sporangium wall is composed of a layer of cells. The capsule wall has a longitudinal tangential wall and lateral wall thickened cell, called a ring, among which a few cells without thickening are called labial cells, which can cause the sporangium to split and release spores.

5.2 Prothallium (gametophyte)　Microscopically, the prothallium is small, thin, green, slightly heart-shaped thallus with dorsal and ventral surfaces. There are rhizoid on the ventral surface, spermatogonia near the rhizoid, and several archegonia in the heart-shaped thallus.

5.3 The dehydrated specimens

5.3.1 The dehydrated specimens of *Selaginella tamariscina* (P. Beauv.)Spring.. The main stem is long, the root system is dense and forming treelike trunk. The branchlets are clustered in the top of the main stem. Sporophyll in the apex of stem forms apex sporangium

5.3.2 The dehydrated plant specimens of *Lygodium japonicum* (Thunb.)Sw.. Rhizome is bifurcate, covered with black-brown hairs. Leaves are three to four pinnate, rachis glabrous. Spores are fine, tuberculate or verrucous.

Experimental report

1. Write the main characters of algae, fungi, lichens, bryophyta and pteridophyta.
2. List the scientific names and families of the the medicinal plants used in the experiment.

Questions

1. What types of fern leaves can be divided into in terms of source and function?
2. Summarize the pharmaceutical standards of China for the inclusion of medicinal algae, fungi and ferns.

实验七 裸子植物门

目的要求

1. **掌握** 裸子植物的主要特征。
2. **熟悉** 裸子植物门的主要分纲和代表植物。
3. **辨识** 常用的药用裸子植物。

实验材料

带雄球花（小孢子叶球）、雌球花（大孢子叶球）和球果的油松枝条；银杏带种子的枝条；侧柏腊叶标本；草麻黄腊叶标本等。

实验用品

解剖镜、放大镜、解剖器具等。

内容与方法

1. 观察油松和侧柏。

（1）取油松的雄球花，观察螺旋状排列的雄蕊。用镊子取一枚雄蕊，解剖镜下观察花粉囊的形状。

（2）取油松的雌球花，观察螺旋状排列的珠鳞。用刀片剥取一枚珠鳞可见到背面基部生一片小苞鳞，腹面基部着生2枚胚珠。

（3）取侧柏腊叶标本，观察枝叶形态、雄球花和雌球花结构等。

2. 观察银杏标本，长枝和短枝之分。观察叶的形态、脉序和种子的三层种皮特征。

3. 取草麻黄标本，观察枝叶形态、球果结构等。

4. 观察校园里的裸子植物，记录典型特征。

实验报告

1. 绘制油松雌球花纵切面简图。

2. 列表记录实验室和校园中观察到的裸子植物的学名、科名。

思考题

裸子植物的主要特征有哪些？

Experiment 7   Gymnospermae

Aims and demands

1. Grasp the main characters of gymnospermae.
2. Familiarize with the division of gymnospermae and its representative plants.
3. Identify commonly medicinal gymnosperms.

Experimental materials

1. The branches of Yousong (*Pinus tabuliformis* Carr.) with microstrobilus and ovulate strobilus branches of Yinxing (*Ginkgo biloba* L.) with seeds.

2. Dehydrated plant specimens of Cebai (*Platycladus orientalis* (L.) Franco) and Caomahuang (*Ephedra sinica* Stapf.).

Experimental Appliances

Anatomic microscope, magnifying glass and dissectors, etc.

Contents and Procedures

1. Observe *Pinus tabuliformis* and *Platycladus orientalis.*

1.1 Take microstrobilus of *Pinus tabulaeformis* to observe the spiral arrangement of stamens, and then observe its shape of anther sac under anatomic microscope.

1.2 Take ovulate strobilus of *Pinus tabulaeformis*, to observe the spiral arrangement of pearl scale. Using a blade to peel pearl scale, observing the abaxial base of a small bract scale, the ventral base of 2 ovules.

1.3 Observe the characters of *Platycladus orientalis* (L.) Franco, such as the shape of leaves and structure of microsporangia or macrosporophylls.

2. Observe the fresh branches of *Ginkgo biloba,* note long shoot and short shoot. Observe its leaf morphology, venation and three layers of seed coat.

3. Take specimens of *Ephedra sinica* and observe its morphology and cone structure of branches and leaves.

4. Observe gymnosperms on campus and record typical features.

Experiment report

1. Draw vertical section of the microstrobilus of *Pinus tabuliformis*.
2. List scientific and family names of the gymnospermae observed in the laboratory and on campus.

Questions

What are the main characteristics of gymnosperms?

实验八 | 桑科、蓼科、毛茛科、木兰科、十字花科

PPT

目的要求

1. **掌握** 桑科、蓼科、毛茛科、木兰科、十字花科植物的主要特征。
2. **熟悉** 植物各器官特征的分类学描述方法；花的解剖方法并能写出花程式。
3. **识别** 桑科、蓼科、毛茛科、木兰科、十字花科的主要药用植物；学会使用植物分类检索表，并能编写简单的科、属或种的检索表。

实验材料

1. 桑科
（1）代表植物 桑枝条，花序及果序标本。
（2）其他药用植物 大麻植株及花果。

2. 蓼科
（1）代表植物 何首乌植株及花果。
（2）其他药用植物 药用大黄植株及花果；虎杖植株及花果。

3. 毛茛科
（1）代表植物 北乌头植株及花果。
（2）其他药用植物 黄连植株及花果；升麻植株及花果。

4. 木兰科
（1）代表植物 五味子植株及花果。
（2）其他药用植物 南五味子植株及花果；厚朴枝条及花果。

5. 十字花科
（1）代表植物 菘蓝或萝卜植株及花果。
（2）其他药用植物 芸苔（油菜）植株及花果；白芥或独行菜植株及花果。

仪器用品

放大镜、解剖镜、解剖用具。

内容与方法

（一）观察解剖各代表植物
1. 桑科
桑 取桑的雌、雄枝条分别观察。观察叶片、花序、花的形态。注意雄花的花被片、雄蕊的形态、类型和数目。注意雌花的雌蕊有无花柱、柱头是否2裂、子房的位置等。识别果实类型并

判断食用部分是花的哪部分结构变化来的。

2. 蓼科

何首乌 取带花或果的何首乌植物。观察植株、叶片、花序、花或果实的形态。注意块根的形状和颜色。注意脉序类型，注意叶片是否有托叶鞘。注意花的颜色、花被片数目和轮数，外轮花被片是否背部有翅。识别果实类型，注意果实是否有3棱。

3. 毛茛科

北乌头 取带花或果的北乌头植物。观察植株、叶片、花序或果实的形态。注意块根的形状。注意叶片是否3全裂。注意花序类型。注意萼片数目、颜色，上萼片是否呈盔状，最下花被是否有长爪。注意雄蕊数目。注意心皮数目和是否离生。识别果实类型。

4. 木兰科

北五味子 取带花或果实的北五味子枝条。观察叶片、花或果实的形态。注意茎的质地和缠绕性。注意花的类型，是否为单性花？注意花被片、雄蕊的形状、类型和数目。注意雌蕊心皮的数目，判断是否为离生单雌蕊。识别果实类型。

5. 十字花科

菘蓝或萝卜 取带花或果实的菘蓝或萝卜植物。观察植株、叶片、花序、花或果实的形态。注意花萼、花瓣、雄蕊的形态、类型、数目。注意雌蕊心皮的数目、假隔膜及子房室数，判断胎座类型。识别果实类型。

（二）观察其他药用植物标本

1. 桑科

大麻 取带花或果的大麻植物。观察植株、叶片、花或果实的形态。注意叶序，叶片是否掌状全裂。注意花是否为单性异株。注意雄花和雌花的花序。注意雌花的苞片和花被片。识别果实类型。

2. 蓼科

（1）药用大黄 取带花或果的药用大黄植物。观察植株、叶片、花序、花或果实的形态。注意根和根茎的断面，是否有星点构造。注意叶裂是否为掌状浅裂，托叶鞘是否筒状。注意花被片、雄蕊和雌蕊的形态、类型和数目。识别果实类型，注意果实是否具3棱，是否沿棱生翅。

（2）虎杖 取带花或果的虎杖植物。观察植株、叶片、花序、花或果实的形态。注意根状茎的特征。注意茎上是否有紫红色斑点，节间是否明显。注意雄花的花被片、雄蕊的形态和数目。注意雌花的花被片、花柱和柱头的形态和数目。识别果实类型，注意是否具三棱，是否有宿存的花被片。

3. 毛茛科

（1）黄连 取带花或果的黄连植物。观察植株、叶片、花序、花或果实的形态。注意苞片、花萼、花瓣、雄蕊和雌蕊的形态和数目。识别果实类型。注意根状茎颜色是否黄色，分枝是否成簇状。注意叶是否3全裂，中央裂片具细柄，侧生裂片不等2深裂，边缘有尖锐锯齿。

（2）升麻 取带花或果的升麻植物。观察植株、叶片、花序、花或果实的形态。注意花萼、花瓣、雄蕊和雌蕊的形态和数目。识别果实类型，是否为聚合蓇葖果。注意根状茎是否粗壮，是否具有多数内陷的圆洞状老茎残迹。注意复叶是几回羽状复叶，小叶片是否卵形或菱形。注意花序是否为圆锥花序且顶生。

4. 木兰科

（1）南五味子 取带花或果的南五味子枝条。观察茎、叶片、花或果实的形态。注意花被片、雄蕊和雌蕊的形态和数目。识别果实类型，是否为聚合浆果。注意叶是否近革质，椭圆形，叶缘是否可见稀疏锯齿。注意花是否单性异株，花托果时是否不延长。注意与北五味子和南五味子的区别。

（2）厚朴　取带花或果的厚朴枝条。观察茎、叶片、花或果实的形态。注意花被片、雄蕊和雌蕊的形态和数目。识别果实类型，是否为聚合蓇葖果。注意叶是否革质，倒卵形，集生枝顶。注意花是否顶生，大，白色，厚肉质。

5. 十字花科

（1）芸苔（油菜）　取带花或果的芸苔植物。观察植株、叶片、花序、花或果实的形态、类型。注意花萼、花瓣、雄蕊的形态、类型、数目。注意雌蕊心皮的数目、假隔膜及子房室数，判断胎座类型。识别果实类型，是否为长角果。注意基生叶是否大头羽裂，基部抱茎；下部茎生叶是否羽状半裂，基部抱茎；上部茎生叶是否长圆状倒卵形、长圆形或长圆状披针形，基部心形抱茎，两侧有垂耳。注意花序是否为总状花序。

（2）白芥或独行菜　取带花或果的白芥或独行菜植物。观察植株、叶片、花序、花或果实的形态、类型。注意花萼、花瓣、雄蕊的形态、类型、数目。注意雌蕊心皮的数目、假隔膜及子房室数，判断胎座类型。识别果实类型，是否为长角果。注意花序是否为总状花序。

实验报告

1. 写出以上五个科植物的主要特征及代表药用植物。
2. 写出上述观察药用植物五个科的花程式。
3. 绘厚朴花的解剖图，并注明各部分名称。
4. 在观察过程中学习使用植物分类检索表，并选择一种植物写出其检索路线。

思考题

1. 蓼属、酸模属、大黄属有何区别？
2. 比较毛茛科中乌头属、毛茛属、铁线莲属的主要异同点，并各列出2种药用植物。
3. 为什么说木兰科是现存被子植物中最原始类群？
4. 比较毛茛科与木兰科的主要异同点。

习题

Experiment 8 — Moraceae, Polyganaceae, Ranunculaceae, Magnoliaceae, Cruciferae

Aim and demand

1. Grasp the main morphological characters of Moraceae, Polygonaceae, Ranunculaceae, Magnoliaceae and Cruciferae.

2. Familiarize with the description of plant morphology and anatomical method. Familiarize with the anatomy of flower and write the floral formula.

3. Recognize the main medicinal plants of the above families. Learn how to use the plant classification keys to identify plants, and write simple classification keys to families, genera or species.

Experiment materials

1. Moraceae
1.1 Typical plant: the plant with branch, inflorescences and infructescence of Sang (*Morus alba* L.).

1.2 Other medicinal plants: the herb with flower of Dama (*Cannabis sativa* L.).

2. Polygonaceae
2.1 Typical plant: the herb with flower and fruit of Heshouwu (*Fallopia multiflora* (Thunb.) Harald).

2.2 Other medicinal plants: the herb with flower and fruit of Yaoyongdahuang (*Rheum officinale* Baill.) and Huzhang (*Reynoutria japonica* Houtt.).

3. Ranunculaceae
3.1 Typical plant: the herb with flower and fruit of Beiwutou (*Aconitum kusnezoffii* Reichb.)

3.2 Other medicinal plants: the herb with flower and fruit of Huanglian (*Coptis chinensis* Franch.); the herb with flower and fruit of Shengma (*Cimicifuga foetida* L.).

4. Magnoliaceae
4.1 Typical plant: the herb with flower and fruit of Wuweizi (*Schisandra chinensis* (Turcz.) Baill.).

4.2 Other medicinal plants: the herb with flower or fruit of Nanwuweizi (*Kadsura longipedunculata* Finet et Gagnep.) and Houpo (*Magnolia officinalis* Rehd.et Wils.).

5. Cruciferae
5.1 Typical plant: the herb with flower and fruit of Songlan (*Isatis indigotica* Fort.) and Luobo (*Raphanus sativus* L.).

5.2 Other medicinal plants: the herb with flower and fruit of Yuntai (*Brassica campestris* L.), Baijie (*Sinapis alba* L.) or Duxingcai (*Lepidium apetalum* Willd.).

Instruments and appliances

Magnifying lens, anatomical lens, dissector.

Contents and Procedures

1. Observe and dissect the typical plants

1.1 Moraceae

Morus alba The female and male branches of *Morus alba* are observed respectively. Observe the morphology of leaves, inflorescence and flowers. Note the shape, type and number of the perianths and stamen of the male flower. Pay attention to whether the pistil of the female flower has a style, whether the stigma is 2-lobed, the position of the ovary, etc. Identify fruit type and determine which part of the flower did the edible part is from.

1.2 Polygonaceae

Fallopia multiflora Take the plant of *Fallopia multiflorarum* with flower or fruit. Observe the morphology of plants, leaves, inflorescence, flowers or fruits. Note the shape and color of the root tuber. Note venation and ocrea. Note the color of the flowers, the number of tepals and the number of whorls, and whether the outer whorl have wings on the back. Identify the fruit type and note whether the fruit is trigonous.

1.3 Ranunculaceae

Aconitum kunezoffii Take the plant of *Aconitum* kunezoffii with flower or fruit. Observe the morphology of plants, leaves, inflorescence or fruits. Notice the shape of the root tuber. Pay attention to whether the leaf blade is 3-sect. Note the inflorescence type. Note the number and color of sepals, whether the upper sepal is helmet-shaped, and whether the lower petal has long claw. Note the number of stamens. Note the number of carpels and whether they are free. Identify fruit type.

1.4 Magnoliaceae

Schisandra chinensis Take the branch of *Schisandra chinensis* with flower or fruit. Observe the morphology of leaves, flowers or fruits. Pay attention to the texture and twining of the stem. Note the type of flower, weather it is unisexual or not. Note the shape, type, and number of tepals and stamens. Note the number of pistil carpels and determine whether it is a free single pistil. Identify fruit type.

1.5 Cruciferae

Isatis indigotica or *Raphanus sativus* Take the plant of *Isatis indigotica* or *Raphanus sativus* with flower or fruit. Observe the morphology of plants, leaves, inflorescence, flowers or fruits. Note the shape, type and number of calyx, petals and stamens. Note the number of pistil carpel, pseudoseptum, ovary locule and the type of placenta. Identify fruit type. The root and leaf of *Isatis indigotica* are used as medicine.

2. Observe other medicinal plants

2.1 Moraceae

Cannabis sativa Take the plant of *Cannabis sativa* with flower or fruit. Observe the morphology of plants, leaves, flowers or fruits. Note phyllotaxy and whether the leaves are palmately. Note if the flowers are unisexual. Note male and female the inflorescence. Pay attention to the bracts and tepals of the female flowers. Identify fruit type.

2.2 Polygonaceae

2.2.1 *Rheum officinale* Take the plant of *Rheum officinale* with flower or fruit. Observe the morphology of plants, leaves, inflorescence, flowers or fruits. Note the cross section of the root and rhizome, and see if there are star-point structures. Note whether the leaf cleft is palmately lobed and the ocrea is tubular. Note the shape, type and number of tepals, stamens and pistil. Identify the type of fruit, pay attention to whether the fruit has 3 arrises and wings along the arrises.

2.2.2 *Reynoutria japonica* Take the plant of *Reynoutria japonica* with flower or fruit. Observe the morphology of plants, leaves, inflorescence, flowers or fruits. Note the characteristics of the rhizome. Pay attention to whether there are purple spots on the stem and whether the internode is obvious. Note the shape and number of the tepals and stamens of the male flowers. Note the shape and number of tepals, styles, and stigmas of the female flowers. Identify fruit type, pay attention to whether the fruits are trigonous and have persistent tepals.

2.3 Ranunculaceae

2.3.1 *Coptis chinensis* Take the plant of *Coptis chinensis* with flower or fruit. Observe the morphology of plants, leaves, inflorescence, flowers or fruits. Note the shape and number of bracts, calyx, petals, stamens, and pistil. Identify fruit type. Note if the rhizome is yellow and the branches are clustered. Pay attention to whether the leaves are 3-lobed, the centrol lobes has slender stipe, while the lateral lobesis unequal 2-deep with sharp serrate on the edge.

2.3.2 *Cimicifuga foetida* Take the plant of *Cimicifuga foetida* with flower or fruit. Observe the morphology of plants, leaves, inflorescence, flowers or fruits. Note the shape and number of calyx, petals, stamens, and pistil. Identify fruit type, whether it is a aggregated follicle. Note whether the rhizome is thickened and has a large number of hollow round old stem remnants. Note whether the compound leaves are several pinnate compound leaves, and whether the leaflets are ovate or rhombic. Note whether the inflorescences are panicles and terminal.

2.4 Magnoliaceae

2.4.1 *Kadsura longipedunculata* Take the branch of *Kadsura longipedunculata* with flower or fruit. Observe the morphology of stem, leaves, flowers or fruits. Note the shape and number of tepals, stamens, and pistil. Identify fruit type, whether it is a aggregate berry. Note whether the leaves are nearly leathery, elliptic, and sparsely serrate on the edges. Note whether the flowers are unisexual and dioecism, receptacle not prolonged while making a fruit. Note the difference with *Kadsura longipedunculata* and *Schisandra chinensis*.

2.4.2 *Magnolia officinalis* Take the branch of *Magnolia officinalis* with flower or fruit. Observe the morphology of stem, leaves, flowers or fruits. Note the shape and number of tepals, stamens, and pistil. Identify fruit type, whether it is a aggregated follicle. Note if leaves are leathery, obovate, set at apex. Note if the flowers are terminal, large, white, thick and fleshy.

2.5 Cruciferae

2.5.1 *Brassica campestris* Take the plant of *Brassica campestris* with flower or fruit. Observe the morphology of plants, leaves, inflorescence, flowers or fruits. Note the shape, type and number of calyx, petals and stamens. Note the number of pistil carpel, pseudoseptum and ovary chamber and determine the type of placenta. Identify the type of fruit and whether it is a silique. Note whether the basal leaves are big-headed divided, basally clasping stem; lower cauline leaves pinnately divided, basally clasping stem; upper cauline leaves oblong-obovate, oblong or oblong-lanceolate, base cordate, with lobes on both sides. Note if the inflorescence is raceme.

2.5.2 *Sinapis alba* or *Lepidium apetalum* Take the plant of *Sinapis alba* or *Lepidium apetalum* with flower or fruit. Observe the morphology of plants, leaves, inflorescence, flowers or fruits. Note the shape, type and number of calyx, petals and stamens. Note the number of pistil carpel, pseudoseptum and ovary chamber and determine the type of placenta. Identify the type of fruit and whether it is a silique or silicle. Note if the inflorescence is raceme.

Experiment report

1. Write the main characters of Moraceae,Polygonaceae, Ranunculaceae, Cruciferae, Magnoliaceae and commonly used medicinal plants.

2. Write the flower formula of the above five families.

3. Draw the anatomy of the flower of *Magnolia officinalis* and note the name of each part.

4. Learn to use the plant classification key during the observation, and choose a plant to describe the identification process.

Questions

1. What are the differences between *Polygonum*, *Rumex* and *Rheum*?

2. Describe the main differences and similarities among *Ranunculus*, *Aconitum* and *Clematis* of Ranunculaceae, and list two pharmaceutical plants of every genus.

3. Why Magnoliaceae is the most primitive group of the extant angiosperms?

4. Compare the main differences and similarities between Ranunculaceae with Magnoliaceae.

实验九 蔷薇科、豆科、芸香科、大戟科、伞形科

PPT

目的要求

1. **掌握** 蔷薇科、豆科、芸香科、大戟科、伞形科等科的主要形态特征。
2. **熟悉** 对植物各器官特征的分类学描述方法；花的解剖方法并能写出花程式；学会使用植物分类检索表，并能编写简单的科、属或种的检索表。
3. **识别** 蔷薇科、豆科、芸香科、大戟科、伞形科的主要药用植物。

实验材料

（一）蔷薇科
1. 代表植物
（1）绣线菊亚科　绣线菊带花果的枝条。
（2）蔷薇亚科　龙芽草带花果的植株。
（3）梅亚科　杏带花果的枝条。
（4）梨亚科　山楂带花果的枝条。
2. 其他药用植物　贴梗海棠带花枝条、果实。
（二）豆科
1. 代表植物
（1）含羞草亚科　合欢带花果的枝条。
（2）云实亚科（苏木亚科）　紫荆带花果的枝条。
（3）蝶形花亚科　槐带花果的枝条。
2. 其他药用植物　决明、甘草等植物的带花果的枝条、植株。
（三）芸香科
1. 代表植物　柑橘带花果的枝条。
2. 其他药用植物　柚带花果的枝条。
（四）大戟科
1. 代表植物　大戟带花果的植株。
2. 其他药用植物　蓖麻带花果的植株。
（五）伞形科
1. 代表植物　当归或野胡萝卜带花果的植株。
2. 其他药用植物　前胡带花果的植株。

仪器用品

放大镜、解剖镜、解剖用具。

内容与方法

（一）观察解剖各代表植物

1. 蔷薇科

（1）绣线菊亚科

绣线菊　取带花和果的绣线菊的枝条。观察枝条、叶片、花序、花或果实的形态。注意叶序、叶形，观察有无托叶，属何种花序类型，取一朵小花放在放大镜下观察萼片、花瓣、雄蕊、雌蕊的数目。注意花筒的形状及心皮的形态和数目，观察花是周位花还是下位花。识别果实类型。

（2）蔷薇亚科

龙芽草　取带花和果的龙芽草植株。观察叶形、叶序、花序、花或果实的形态。注意小叶的形状及大小的不同，大小相间排列。顶生总状花序。取花观察，花筒顶端有无一圈钩状刚毛，横切子房观察心皮数目，并观察果实类型。

（3）梅亚科

杏　取带花和果的杏枝条。注意叶形，观察有无托叶，取一朵小花观察萼片、花瓣、雄蕊的数目。纵剖花，注意花筒的形状，花被及雄蕊着生在花筒边缘。雌蕊一枚，着生在花筒底部。观察果实类型。

（4）梨亚科

山楂　取带花和果的山楂枝条观察。观察枝条、叶片、花序、花或果实的形态。注意叶序、叶形，观察托叶形状。观察花序类型。取一朵小花放在放大镜下观察花柱特征及子房位置，横切子房，注意由几个心皮合生成。识别胎座类型，取果实横切，观察梨果特点。

2. 豆科

（1）含羞草亚科

合欢　取带花和果的合欢枝条。观察枝条、叶片、花序、花或果实的形态。注意花的对称性以及花萼、花冠的连合情况，雄蕊和心皮的数目。观察果实类型。

（2）云实亚科（苏木亚科）

紫荆　取带花和果的紫荆枝条。观察枝条、叶片、花序、花或果实的形态。注意其树皮和小枝灰白色。识别花冠的结构、颜色、类型。

（3）蝶形花亚科

槐　取带花和果的槐树的枝条。观察枝条、叶片、花序、花或果实的形态。注意观察花冠的类型和颜色。观察旗瓣、龙骨瓣和翼瓣的位置关系和大小；注意雄蕊数目及联合还是分离；果实和种子形状。

3. 芸香科

柑橘　取带花和果的柑橘的枝条。观察枝条、叶片、花序、花或果实的形态。解剖花和果实并观察内部结构。

4. 大戟科

大戟　取带花和果的大戟植株。观察植株、叶片、花序、花或果实的形态。重点观察其花序类型和结构，注意总苞片、腺体、雄花和雌花形态和数量。识别果实类型。

5. 伞形科

当归或野胡萝卜　取带花和果的当归或野胡萝卜植株。观察植株、叶片、花序、花或果实的形态。注意观察复叶类型和叶柄基部性状。观察花序类型，注意总苞片和小苞片的特征。注意果实的类型。

（二）观察其他药用植物标本

1. 蔷薇科

贴梗海棠　取带花和果的贴梗海棠枝条。注意观察枝是否有刺，观察叶片、花序、花和果实的形状和类型。

2. 豆科

（1）决明　取带花和果的决明植株。观察植株、叶片、花序、花或果实的形态。注意其小叶的形状和质地；注意观察叶轴上最下方一对小叶间有棒状腺体1枚；托叶线状，被柔毛。总状花序；花黄色荚果，种子长菱形。

（2）甘草　取带花和果的甘草植株。观察植株、叶片、花序、花或果实的形态。注意观察根与根状茎的形态和颜色。注意托叶三角状披针形，两面密被白色短柔毛；叶柄密被褐色腺点和短柔毛。注意花的数量、排列方式；子房密被刺毛状腺体。注意荚果形态。

3. 芸香科

柚　取带花和果的柚枝条。观察植株、叶片、花序、花或果实的形态。注意观察嫩枝、叶背、花梗、花萼及子房均被柔毛。观察复叶的类型，小叶的形态和数目，观察花序类型。解剖果实结构，注意其油点大而凸起。

4. 大戟科

蓖麻　取带花和果的蓖麻植株。观察植株、叶片、花序、花或果实的形态。注意观察其叶典型的掌状分裂；叶柄粗壮，中空，顶端具2枚盘状腺体。注意观察花序及花部各部分结构及苞片形状；单性花；观察果实类型和特征；种子种阜较大。

5. 伞形科

前胡　取带花和果的前胡植株。观察植株、叶片、花序、花或果实的形态。注意观察其根为圆锥状柱形；叶为二至三回三出式羽状分裂，叶裂片边缘具粗锯齿或圆锯齿。注意观察花序的类型和结构。解剖花，并观察花的各部位颜色、数量和形态，花白色。

实验报告

1. 认真观察各科植物，列举以上各科常见药用植物，总结说明关键性鉴别特征。
2. 比较豆科三个亚科的主要特征和区别。
3. 写出观察的植物花程式。
4. 编制所观察植物的检索表。

思考题

1. 各个科最典型的鉴别特征表现在哪些方面？如何利用这些特征来鉴别植物？
2. 根据哪些特征在分类学研究中把豆科划分为三个亚科？

Experiment 9 | Rosaceae, Leguminosae, Rutaceae, Euphorbiaceae, Umbelliferae

Aim and demand

1. Grasp the main morphological characteristics of Rosaceae, Leguminosae, Rutaceae, Euphorbiaceae and Umbelliferae.

2. Familiarize with the description of plant morphology and anatomical method. Familiar with the anatomy of flower and write the floral formula. Learn to use the plant classification keys to identify plants, and write simple classification keys to families, genera or species.

3. Recognize the main medicinal plants of the above five families.

Experiment materials

1. Rosaceae

1.1 Typical plant The plants listed below of four subfamilies of Rosaceae with flower and fruit.

1.1.1 Spiraecideae Xiuxianju (*Spiraea salicifolia* L.).

1.1.2 Rosoideae Longyacao (*Agrimonia pilosa* Ledeb.).

1.1.3 Prunoideae Xing (*Armeniaca vulgaris* Lam.)

1.1.4 Maloideae Shanzha (*Crataegus pinnatifida* Bge.).

1.2 Other medicinal plants Tiegenghaitang (*Chaenomeles speciosa* (Sweet) Nakai.).

2. Leguminosae

2.1 Typical plant the plants listed below of three subfamilies of Leguminosae with flower and fruit.

Mimosoideae: Hehuan (*Albizia julibrissin* Durazz.).

Caesalpinoideae: Zijing (*Cercis chinensis* Bunge.)

Papilionoideae: Huai (*Sophora japonica* Linn.)

2.2 Other medicinal plants The branch of Jueming (*Cassia tora* L.) and Gancao (*Glycyrrhiza uralensis* Fisch.) with flower and fruit.

3. Rutaceae

3.1 Typical plant The branch of Ganju (*Citrus reticulata* Blanco) with flower or fruit.

3.2 Other medicinal plants The branches of You (*Citrus maxima* (Burman)) Merrill with flower or fruit.

4. Euphorbiaceae

4.1 Typical plant The plant of Daji (*Euphorbia pekinensis* Rupr.) with flower or fruit.

4.2 Other medicinal plants The plant of Bima (*Ricinus communis* Linn.) with flowers or fruits.

5. Umbelliferae

5.1 Typical plant The plant of Danggui (*Angelica sinensis* (Oliv.) Diels) and Yehuluobo (*Daucus* *carota* Linn.) with flowers or fruits.

5.2 Other medicinal plants The plant of Qianhu (*Peucedanum praeruptorum* Dunn) with flower or fruit.

Instruments and appliances

Magnifier, anatomical lens, dissector.

Contents and Procedures

1. Dissect and observe the typical plants of every family

1.1 Rosaceae

1.1.1 Spiraeoideae

Spiraea salicifolia Take the shoot of *Spiraea salicifolia* with flower or fruit. Observe the phyllotaxy, phylliform and inflorescence; and if there are stipules or not. Observe carefully the type of inflorescence. Pick up a single little flower and observe with a magnifier to count the number of sepals, petals, stamens and pistils. Pay attention to the shape of hypanthium and number of the carpel. Is it a peripheral flower or inferior flower? Recognize the type of fruit.

1.1.2 Rosoideae

Agrimonia pilosa Take the plant of *Agrimonia pilosa* with flower or fruit. Observe the phyllotaxy, phylliform, inflorescence, the shape of flower and fruit. Pay attention to the shape and size of leaflets, little alternating with large. Terminal racemes. Observe whether there is a ring of hooked hair at the top of hypanthium, and count the number of carpels. Observe the type of fruit.

1.1.3 Prunoideae

Armeniaca vulgaris Take the branch of *Armeniaca vulgaris* with flower or fruit. Observe the phylliform, pay attention to the shape of leaves, if there is stipule or not count the number of sepals, petals, stamens. Dissect the flower longitudinally and pay attention to the shape of hypanthium. Perianth and stamen are at the edge of hypanthium. One pistil is at the bottom of hypanthium. Observe the fruit type.

1.1.4 Maloideae

Crataegus pinnatifida Take the shoot of *Crataegus pinnatifida* with flower or fruit. Observe the phyllotaxy, phylliform, inflorescence and the type of flower and fruit. Pay attention to the position of ovary and character of style. Cut ovary transversely, to observe the number of carpels, type of placenta and characteristics of pome.

1.2 Leguminosae

1.2.1 Mimosoideae

Albizia julibrissin Take the branch of *Albizia julibrissin* with flower or fruit. Observe the morphology of branches, leaves, inflorescences, flowers or fruits. Note the symmetry of the flowers and the union of calyx and corolla, the number of stamens and carpel. Note the fruit type.

1.2.2 Caesalpinoideae

Cercis chinensis Take the branch of *Cercis chinensis* with flower or fruit. Observe the morphology of branches, leaves, inflorescences, flowers or fruits. Pay attention to the color of the bark and branchlets. Identify the structure, color and type of corolla.

1.2.3 Papilionoideae

Sophora japonica   Take the branch of *Sophora japonica* with flower or fruit. Observe the morphology of branches, leaves, inflorescences, flowers or fruits. Note the color and type of corolla. Observe the position and size of claw, wing and keel. Note the number of stamens and union or separation. Note the shape of fruit and seed.

1.3 Rutaceae

Citrus reticulata   Take the branch of *Citrus reticulata* with flower or fruit. Observe the morphology of branches, leaves, inflorescences, flowers or fruits. Pay attention to the types of leaves, the morphology and texture of the leaves. Dissect a flower and a fruit to observe its internal structure.

1.4 Euphorbiaceae

Euphorbia pekinensis   Take the branch of *Euphorbia pekinensis* with flower or fruit. Observe the morphology of branches, leaves, inflorescences, flowers or fruits. Pay attention to the shape and number of involucral bracts, glands, male and female flowers. Note the fruit type.

1.5 Umbelliferae

Angelica sinensis or *Daucus carota*   Take the plant of *Angelica sinensis* or *Daucus carota* with flower or fruit. Observe the morphology of plants, leaves, flowers or fruits. Note the types of compound leaves and the characteristics of petiole base. Observe the type of inflorescences, the morphology of involucre and bracteoles. Observe the shape of fruit.

2.　Observe other samples of medicinal plants

2.1 Rosaceae

Chaenomeles speciosa   Take the shoot of *Chaenomeles speciosa* with flower or fruit. Note whether the branches have thorns. Observe the shape and type of leaves, inflorescence, flowers or fruits.

2.2 Leguminosae

2.2.1 *Cassia tora*   Take the plant of *Cassia tora* with flower or fruit. Pay attention to the branches, leaves, flowers or fruits. Note the shape and texture of leaflets, rachis with a club-shaped gland between lowest leaflets. Stipules caducous, linear, racemes, petals yellow legume, seeds rhomboid.

2.2.2 *Glycyrrhiza uralensis*   Take the plant of *Glycyrrhiza uralensis* with flower and fruit. Observe the morphology of plants, leaves, inflorescence, flowers or fruits. Note the color and shape of root and rhizome. Stipules triangular-lanceolate, densely white pubescent. Petiole densely brown glandular punctate and pubescent. Pay attention to the number and arrangement of the flowers. Ovary densely thorn hairy glandular. Pay attention to the characteristics of the legume.

2.3 Rutaceae

Citrus maxima   Take the branch of *Citrus maxima* with flower or fruit. Observe the morphology of plants, leaves, flowers or fruits. Note that the shoots, abaxial surface of leaves, peduncles, calyx and ovaries are pilose. Observe the types of compound leaves, the number and shape of leaflets, the type of inflorescence. Dissect a fruit to observe the each part of the fruit with large prominent oil dots.

2.4 Euphorbiaceae

Ricinus communis   Take the plant of *Ricinus communis* with flower and fruit. Observe the morphology of plants, leaves, inflorescence, flowers or fruits. Observe the typical palmate division of the leaves, with petiole stout and hollow, with 2 disk-shaped glands at apex. Note the type of inflorescence, the structure of flower, the shape of bracts. The flowers are unisexual. Observe fruit types and characteristics, seeds caruncle obvious.

2.5 Umbelliferae

Peucedanum praeruptorum　　Take the plant of *Peucedanum praeruptorum* with flower and fruit. Observe the morphology of plants, leaves, inflorescence, flowers or fruits. The roots are observed to be conical and columnar. Leaves 2-3-ternate-pinnate, leaf lobes margin with coarse serrated or circular serrated. Observe the type and structure of inflorescence. Dissect the flower and observe its color, number and shape; petal white.

💬 Experiment report

1. List the main characteristics of the given families and commonly used medicinal plants by observation.

2. Compare the main characteristics and differences of the three subfamilies of Leguminosae.

3. Write the plant flower formula of observed plants.

4. Compile the Plant Classification Key of observed plants.

Questions

1. What are the typical distinguishing features of each family? How could these characteristics be used to identify plants?

2. Leguminosae is divided into three subfamilies in taxonomic study accordind to which characteristics?

实验十　木犀科、夹竹桃科、唇形科、玄参科、茄科

PPT

微课

目的要求

1. **掌握**　木犀科、夹竹桃科、唇形科、玄参科、茄科的主要形态特征。

2. **熟悉**　对植物各器官特征的分类学描述方法，熟悉花的解剖方法并能写出花程式，学会使用植物分类检索表。

3. **识别**　木犀科、夹竹桃科、唇形科、玄参科、茄科主要的药用植物。

实验材料

（一）木犀科

1. **代表植物**　连翘带花果的枝条。

2. **其他药用植物**　女贞带花果的枝条。

（二）夹竹桃科

1. **代表植物**　络石带花果藤茎。

2. **其他药用植物**　罗布麻带花全株。

（三）唇形科

1. **代表植物**　益母草。

2. **其他药用植物**　丹参和黄芩带花、果的植株。

（四）玄参科

1. **代表植物**　地黄带花、果的全株。

2. **其他药用植物**　玄参带花、果的植株。

（五）茄科

1. **代表植物**　白花曼陀罗带枝、叶、花、果的植株。

2. **其他药用植物**　宁夏枸杞带花果的枝条。

仪器用品

放大镜、解剖镜、解剖用具。

内容与方法

（一）观察解剖下列各代表植物

1. 木犀科

连翘　取带花果的连翘枝条。观察茎的形状，纵剖枝条观察髓是空心还是实心。看清叶的着生方式、叶形后，取花观察其花被、雄蕊、雌蕊、果实、种子数目类型及子房位置，说明由几个

心皮构成及胎座类型。

2. 夹竹桃科

络石　取带花果的络石枝条。观察叶形和叶序；注意是否有乳汁；注意花冠颜色，观察花萼、花冠、花序及果实类型。

3. 唇形科

益母草　取带花果的益母草植株。观察茎的形状及叶序类型；注意茎下部、中部、上部叶的分裂变化。观察小苞片、花萼形状，观察花冠颜色及类型，注意花序、雄蕊群、雌蕊群类型；观察果实类型、颜色；子房室数目。

4. 玄参科

地黄　取带花的地黄植株。观察全体密被灰白色柔毛及腺毛，是根状茎、块茎还是块根、分辨叶的着生方式、叶形、花序。注意花冠的形状、颜色、上下唇裂片的数目、上唇是否反卷、是否为二强雄蕊。横切子房，观察子房室、胚珠的数目及胎座类型。观察果实和种子的类型、颜色。

5. 茄科

白花曼陀罗　取带花的白花曼陀罗全株。观察叶形、叶序类型。与果实对照观察花萼是否宿存；观察花冠颜色和类型、雄蕊数目，横切子房观察子房室数目。观察果实类型。

（二）观察其他药用植物

1. 木犀科

女贞　取带花果的女贞枝条。观察叶、叶序和花序类型。注意花的各部分基数，观察花冠颜色及类型；横切雌蕊子房，观察心皮、子房室数目；观察果实类型。

2. 夹竹桃科

罗布麻　取带花果的罗布麻枝条。注意叶形和叶序，是否有乳汁，观察花冠和果实类型。

3. 唇形科

（1）丹参　取带花或果实的丹参植株观察。全株密被淡黄色的柔毛及腺毛，注意根的颜色。区别叶序和花序类型、花冠的类型及花冠裂片数；注意雄蕊的形态、类型、数目。注意雌蕊心皮的数目及子房室数。果实类型。

（2）黄芩　取带花果的黄芩植株观察。观察根横断面颜色；注意观察叶的形状及着生的方式；观察花序、花萼和花冠类型，以及花萼上唇背部具盾状物。

4. 玄参科

玄参　取带花果的玄参植株。注意观察根的类型，注意茎的形状，观察叶形、叶序和花序类型；观察花萼、花冠的形状和颜色、雄蕊类型及果实类型。

5. 茄科

宁夏枸杞　取带花果的宁夏枸杞枝条。观察植株、叶、花、果实的形态。注意枝条有无棘刺。注意花冠筒与花冠裂片的长度比例。

💬 **实验报告**

1. 列出上述五个科的主要特征，并写出常用的药用植物。
2. 写出上述所观察植物的花程式。
3. 编写上述观察植物的检索表。

思考题

1. 比较唇形科和玄参科的异同点。
2. 列举与人类关系比较密切的茄科药用植物。

Experiment 10   Oleaceae, Apocynaceae, Labiatae, Scrophulariaceae, Solanaceae

Aim and demand

1. Grasp the main morphological characters of Oleaceae, Apocynaceae, Labiatae, Scrophulariaceae and Solanaceae.

2. Familiarize with the description of plant morphology and anatomical method. Familiarize with the anatomy of flower and write the floral formula. Learn to use the plant classification keys to identify plants.

3. Recognize the main medicinal plants of Oleaceae, Apocynaceae, Labiatae, Scrophulariaceae and Solanaceae.

Experiment materials

1. Oleaceae

1.1 Typical plant   the shoot with flower and fruit of Lianqiao (*Forsythia suspensa* (Thunb.) Vahl.)

1.2 Other medicinal plants   the branch with flower and fruit of Nvzhen (*Ligustrum lucidum* Ait.).

2. Apocynaceae

2.1 Typical plant   the shoot with flower and fruit of Luoshi (*Trachelospermum jasminoides* (Lindl.) Lem.).

2.2 Other medicinal plants   the plant with flower of Luobuma (*Apocynum venetum* L.).

3. Labiatae

3.1 Typical plant   the herb with flower of Yimucao (*Leonurus heterophyllus* Sweet).

3.2 Other medicinal plants   the herb of Danshen (*Salvia miltiorrhiza* Bge.) and Huangqin (*Scutellaria baicalensis* Georgi) with flower and fruit.

4. Scrophulariaceae

4.1 Typical plant   the herb with flower and fruit of Dihuang (*Rehmanniaglutinosa* (Gaertn.) Libosch.).

4.2 Other medicinal plants   the plant of Xuanshen (*Scrophularia ningpoensis* Hemsl) with flower and fruit.

5. Solanaceae

5.1 Typical plant   the shoot with leaf, flower and fruit of Baihuamantuoluo (*Datura metel* L.).

5.2 Other medicinal plants   the shoot with fruit of Ningxiagouqi (*Lycium barbarum* L.).

Instruments and appliances

Magnifier, anatomical lens, dissector.

Contents and Procedures

1. Observe and dissect the typical plants

1.1 Oleaceae

Forsythia suspense   Take the branch of *Forsythia suspense* with flower and fruit. Observe the shape of stem and whether the pith is full or hollow. Describe the phyllotaxy and phylliform. Observe the type and number of calyx, petals, stamens, ovary, stigma, fruit and seed. Describe the number of carpels and placenta type.

1.2 Apocynaceae

Trachelospermum jasminoides   Take the branch of *Trachelospermum jasminoides* with flower and fruit. Observe the phylliform and phyllotaxy, whether there is laticifer. Note the color of corolla. Observe the type of calyx, corolla, inflorescence and fruit.

1.3 Labiatae

Leonurus heterophyllus   Take the plant of *Leonurus heterophyllus* with flower and fruit. Observe the type of phyllotaxy and the shape of stem. Pay attention to division of the lower, middle and upper leaves of the stem. Observe the shape of bractle, calyx. Note the type of inflorescence, stamens and pistils. Observe the color and shape of corolla and fruit. The number of ovary locule.

1.4 Scrophulariaceae

Rehmania glutinosa   Take the plant of *Rehmania glutinosa* with flower or fruit. The herb is covered with a lot of grayish white villi and glandular hairs. Decide whether it is rhizome, tuber or root tuber. Distinguish the phyllotaxy, phylliform, inflorescence. Observe the shape and color of corolla, number of upper and lower lips, whether the upper lip is revoluted, if stamen is didynamous or not. Transversely cut the ovary and count the number of ovary locule, ovules and decide the type of placenta. Observe the shape and color of fruit and seed.

1.5 Solanaceae

Datura metel   Take the plant of *Datura metel* with flower or fruit. Describe the phyllotaxy and phylliform. Note the shape of calyx and if it persists. Observe the color and shape of corolla, the number of stamens. Transversely cut the ovary and count the number of ovary locule. Observe the type of fruit.

2. Observe other samples of Medical plants

2.1 Oleaceae

Ligustrum lucidum   Take the branch of *Ligustrum lucidum* with flower or fruit. Observe the phyllotaxy, phylliform and inflorescence. Note the number of calyx, corolla, stamens and pistil. Observe the color and type of corolla. Transversely cut the ovary and count the number of carpel and ovary locule. Observe the type of fruit.

2.2 Apocynaceae

Apocynum venetum   Take the branch of *Apocynum venetum* with flower or fruit. Observe phylliform and phyllotaxy, whether there is laticifer. Pay attention to the type of corolla and fruit.

2.3 Labiatae

2.3.1 *Salvia miltiorrhiza*   Take the plant of *Salvia miltiorrhiza* with flower or fruit. It is densely covered with flavescent vellus hair and glandular hairs. Note the color of root. Discriminate the type of phyllotaxy, inflorescence, corolla and the number of corolla lobes. Note the shape, type and number of stamens. Note the number of pistil carpel and ovary locule. Note the fruit type.

2.3.2 *Scutellaria baicalensis*   Take the plant of *Scutellaria baicalensis* with flower or fruit. Note the color of transverse section of root. Pay attention to the shape and inserted form of leaves. Observe the type of inflorescence, calyx, corolla, and the calyx with a scutellum on the back of the upper lip.

2.4 Scrophulariaceae

Scrophularia ningpoensis   Take the plant of *Scrophularia ningpoensis* with flower or fruit. Note the type of root and the shape of stem. Observe the type of phylliform, phyllotaxy and inflorescence. Observe the color and shape of calyx and corolla. Note the type of stamens and fruit.

2.5 Solanaceae

Lycium barbarum   Take the branch of *Lycium barbarum* with flower or fruit. Observe the morphology of plants, leaves, flowers and fruits. Note the thorns on the branch. Note the length ratio of the corolla tube and corolla lobes.

Experiment report

1. List the main characteristics of the given families and commonly used medicinal plants.
2. Write the plant flower formula of the observed plants.
3. Compile the Plant Classification Key for the observed plants.

Questions

1. Compare the similarities and differences between Labiatae and Scrophulariaceae.
2. List the medicinal plants of Solanaceae closely related to human beings.

实验十一　茜草科、忍冬科、葫芦科、桔梗科、菊科

目的要求

1. **掌握**　茜草科、忍冬科、葫芦科、桔梗科、菊科植物的主要形态特征。
2. **熟悉**　植物各器官特征的分类学描述方法，熟悉花的解剖方法并能写出花程式。
3. **识别**　茜草科、忍冬科、葫芦科、桔梗科、菊科主要的药用植物。

微课

实验材料

（一）茜草科

1. **代表植物**　栀子的枝条、花、果。
2. **其他药用植物**　茜草的枝条、花、果。

（二）忍冬科

1. **代表植物**　忍冬带花茎枝。
2. **其他药用植物**　灰毡毛忍冬带花茎枝；华南忍冬带花茎枝；红腺忍冬带花茎枝。

（三）葫芦科

1. **代表植物**　栝楼带花藤茎、块根、果实、种子。
2. **其他药用植物**　绞股蓝植株及花果。

（四）桔梗科

1. **代表植物**　桔梗植株及花果。
2. **其他药用植物**　沙参带花植株、根；党参带花植株、根。

（五）菊科

1. **管状花亚科**

（1）代表植物　菊花带花全草。
（2）其他药用植物　红花带花序全草；白术带花全草。

2. **舌状花亚科**

（1）代表植物　蒲公英带花全草、果。
（2）其他药用植物　苦苣菜全草。

仪器用品

放大镜、解剖镜、解剖用具。

内容与方法

（一）观察解剖各科代表植物

1. 茜草科

栀子　取带花和果实的栀子。观察植株、叶片、花、果实的形态。注意叶的着生方式和叶形，叶柄内托叶形态。注意花萼、花冠、雄蕊、雌蕊的形态和数目，果实形态。判断子房室数。

2. 忍冬科

忍冬　取带花果的忍冬植物。观察茎枝、叶片、花序、果实的形态。注意幼枝、幼叶形态。注意总花梗、花萼、花瓣的形态。判断花萼是否无毛。

3. 葫芦科

栝楼　分别取栝楼的雌、雄株。分别观察块根、茎、叶、花及果实种子特征。注意根、茎、叶片、花、果实的形态。注意茎卷须、叶片的形态。注意雄花花序，雌花、花冠的形态和类型。识别果实类型。

4. 桔梗科

桔梗　取带花果的桔梗新鲜植株。观察植株、茎、花、果实的形态、类型。注意折断茎叶后是否有乳汁流出。注意花冠的形状，子房位置。识别果实类型。判断子房室数、胎座类型。

5. 菊科

（1）管状花亚科

菊花　取带花菊花植株。观察植株、叶片、花序、果实的形态、类型。折断新鲜植物看是否有乳汁流出。注意小花的形态、类型、性别，是否全为管状花。解剖管状花，判断其是两性花，单性花，还是中性花。判断雄蕊类型。注意雌蕊心皮数目、子房类型及雄蕊。识别果实类型。

（2）舌状花亚科

蒲公英　取带花蒲公英植株。按照菊花的观察方法操作观察，注意其与菊花的区别点。

（二）观察其他药用植物

1. 茜草科

茜草　取带花果的茜草植物。观察根、茎、叶片、花序、果实的形态、类型。注意茎（手摸）、叶片的形态。注意花序、雌蕊心皮数。判断花序类型。

2. 忍冬科

（1）灰毡毛忍冬　取带花茎枝观察，与忍冬比较。注意叶片形态。

（2）华南忍冬　取带花茎枝观察，与忍冬比较。注意萼筒、苞片形态。

（3）红腺忍冬　取带花茎枝观察，与忍冬比较。注意叶片、总花梗形态。

3. 葫芦科

绞股蓝　分别取绞股蓝的雌、雄植株分别。观察茎、叶、花及果实特征。注意茎卷须、复叶、叶片、花序、花冠、果实的形态、类型。识别果实类型。

4. 桔梗科

（1）沙参　取带花果的沙参植物。观察植株、茎、叶、花、果实的形态、类型。注意折断茎叶后有无乳汁。注意花序、花冠的形状、类型。注意花柱基部有杯状花盘、子房位置。识别果实类型。判断子房心皮数、胎座类型。

（2）党参　取带花果的党参植物。观察植株、茎、叶、花、果实。注意叶片、花冠、果实的形态、类型。

5. 菊科

（1）管状花亚科

红花　取带花红花植物。观察叶、花、果实。注意叶片、花序、小花的形态、类型。判断是否全为管状花。

白术　取带花白术植物。观察根状茎、叶、花序。注意花序、花冠的形态、类型。

（2）舌状花亚科

苦荬菜　取带花苦荬菜植物。观察根、茎、叶、花序、小花的形态、类型。注意根、叶片、花序、小花的形态、类型。判断是否全为舌状花。

💬 实验报告

1. 列出以上五个科的主要特征，并列举数种常见药用植物。

2. 写出所观察植物的花程式。

3. 绘菊科药用植物一朵管状两性花、一朵舌状两性花、一朵舌状雌花的纵剖图，并注明各部分名称。

思考题

1. 何为中药"金银花"的来源，如何辨别真假？

2. 简述菊科分成两个亚科的依据是什么？

3. 菊科植物的哪些特征可以说明其进步性，这些特征在适应环境时有什么意义？

习题

Experiment 11 Rubiaceae, Caprifoliaceae, Cucurbitaceae, Campanulaceae, Compositae

Aims and Requirements

1. Grasp the main morphological characteristics of Rubiaceae, Caprifoliaceae, Cucurbitaceae, Campanulaceae and Compositae.

2. Familiarize with the description of plant morphology and anatomical method. Familiarize with the anatomy of flower and write the floral formula.

3. Recognize the main medicinal plants of Rubiaceae, Caprifoliaceae, Cucurbitaceae, Campanulaceae and Compositae.

Experiment materials

1. Rubiaceae

1.1 Typical plant   The shoot with flower, fruit of Zhizi (*Gardenia jasminoides* Ellis).

1.2 Other medicinal plants   The shoot with flower and fruit of Qiancao (*Rubiacordifolia* L.).

2. Caprifoliaceae

2.1 Typical plant   The vines with flower of Rendong (*Lonicera japonica* Thunb.)

2.2 Other medicinal plants   The vines with flower of Huizhanmaorendong (*L. macranthoides* Hand.-Mazz.); the vines with flower of Huananrendong (*L. confuse* DC.); the vines with flower of Hongxianrendong (*L. hypoglauca*Miq.).

3. Cucurbitaceae

3.1 Typical plant   The vines with flower, root tuber, fruit and seed of Gualou (*Trichosanthes kirilowii* Maxim.)

3.2 Other medicinal plants   The stem, leaf, flower and fruit of Jiaogulan (*Gynostemma pentaphyllum* (Thunb.) Makino).

4. Campanulaceae

4.1 Typical plant   The plant with flower and fruit of Jiegeng (*Platycodon grandiflorum* (Jacq.) A. DC.)

4.2 Other medicinal plants   the plant with flower, root of Shashen (*Adenophora stricta* Miq.) and Dangshen (*Codonopsis pilosula* (Franch.) Nannf.).

5. Compositae

5.1 Carduoideae

5.1.1 Typical plant   The plant with flower of Juhua (*Dendranthema morifolium* (Ramat.) Tzvel.).

5.1.2 Other medicinal plants   The plant with inflorescence of Honghua (*Cathamus tinctorius* L.) and Baizhu (*Atratylodes macrocephala* Koidz.).

5.2 Cichorioideae

5.2.1 Typical plant   The plant with flower and fruit of Pugongying (*Taraxacum mongolicum* Hand.-Mazz.)

5.2.2 Other medicinal plants   The plant of Kujucai (*Sonchus oleraceus* L.)

Instrument and appliance

Magnifier, anatomical lens, dissector.

Contents and Procedures

1. Observe and dissect the typical plants

1.1 Rubiaceae

Gardenia jasminoides   Take the plant of *Gardenia jasminoides* with flowers and fruits. Observe the morphology of leaves, flowers and fruits. Pay attention to the phyllotaxy and leaf shape. Note the morphology and number of calyx, corolla, stamen, pistil, and fruit morphology. Determine the number of ovary locule.

1.2 Caprifoliaceae

Lonicera japonica   Take the plant of *Lonicera japonica* with flowers and fruits. Observe the morphology of stem, branch, leaves, inflorescences and fruits. Pay attention to the morphology of young branch and leaves. Note the morphology of the total pedicels, calyx and petals. Determine if the calyx is hairless.

1.3 Cucurbitaceae

Trichosanthes kirilowii   Take the female and male plants of *Trichosanthes kirilowii* with flowers and fruits, respectively. Observe the characteristics of the root, stem, leaves, flowers, fruits and seeds. Pay attention to the morphology of root, stem leaves, flowers, and fruits. Pay attention to the shape of stem tendrils and leaves. Pay attention to the morphology and type of male flowers, female flowers and corollas. Identify fruit type.

1.4 Campanulaceae

Platycodon grandiflorum   Take fresh plant of *Platycodon grandiflorum* with flowers and fruits. Observe the morphology and type of plants, stem, flowers and fruits. Pay attention to whether there is latex flowing out of the broken stem or leaves. Note the shape of the corolla and the location of the ovary. Identify fruit type. Determine the number of ovary locule and placenta type.

1.5 Compositae

1.5.1 Carduoideae

Dendranthema morifolium   Take the plant of *Dendranthema morifolium* with flowers and fruits. Observe the morphology and type of plant, leaves, inflorescences, flowers and fruits. Note the number of pistils, carpels, ovary type, stamens and fruit type. Break the fresh plants to see if there is any latex flowing out. Dissect the small tubular flower to determine whether it is bisexual, unisexual or neutral. Determine the type of stamens. Note the number of pistils, carpels, ovary type and fruit type.

1.5.2 Cichorioideae

Taraxacum mongolicum   Take the plant of *Taraxacum mongolicum* with flowers and fruits. Observe the characteristics of the parts according to the observation method of *Dendranthema morifolium*. Note the differences between them.

2. Observe other samples of medicinal plants

2.1 Rubiaceae

Rubia cordifolia Take plant of *Rubia cordifolia* with flowers and fruits. Observe the morphology and type of root, stem, leaves, inflorescences and fruits. Pay attention to the shape of the stem (touch it by hands) and leaves. Note the number of inflorescences and pistils. Observe the type of inflorescence.

2.2 Caprifoliaceae

2.2.1 *Lonicera macranthoides* Take the vines of *Lonicera macranthoides* with flowers. Observe and compare them with *L. japonica*. Note the leaf shape.

2.2.2 *L. confuse* Take the vines of *L. confuse* with flowers. Observe and compare them with *L. japonica*. Note the calyx tube and bract morphology.

2.2.3 *L. hypoglauca* Take the vines of *L. hypoglauca* with flowers. Observe and compare them with *L. japonica*. Note leaf and pedicel morphology.

2.3 Cucurbitaceae

Gynostemma pentaphyllum Take the female and male plants of *Gynostemma pentaphyllum* with flowers and fruits separately. Observe the characteristics of stem, leaves, flowers and fruits. Note the morphology and type of stem tendrils, compound leaves, leaves, inflorescences, corollas and fruits. Determine fruit type.

2.4 Campanulaceae

2.4.1 *Adenophora stricta* Take the plant of *Adenophora stricta* with flowers and fruits. Observe the morphology of the plants, stem, leaves, flowers and fruits. Pay attention to whether there is latex from the broken stem or leaves. Note the shape and type of the inflorescence and corolla. Note the base of the style has cup-shaped disks and ovary positions. Identify fruit type. Determine the number of carpels and the placenta type.

2.4.2 *Codonopsis pilosula* Take the plant of *Codonopsis pilosula* with flowers and fruits. Observe the morphology of plants, stem, leaves, flowers, and fruits. Pay attention to the shape and type of leaves, corollas and fruits.

2.5 Compositae

2.5.1 Carduoideae

2.5.1.1 *Cathamus tinctorius* Take the plant of *Cathamus tinctorius* with flower. Observe the morphology and type of leaves, flowers and fruits. Pay attention to the shape and type of leaves, inflorescences and floret. Determine if they are all tubular flowers.

2.5.1.2 *Atratylodes macrocephala* Take the plant of *Atratylodes macrocephala* with flowers. Observe the morphology of rhizomes, leaves and inflorescences. Note the shape and type of the inflorescences and corollas.

2.5.2 Cichorioideae

Sonchus oleraceus: Take the plant of *Sonchus oleraceus* with flowers. Observe the morphology root, stem, leaves, inflorescences and floret. Pay attention to the morphology and type of root, leaves, inflorescences and floret. Determine whether the flowers are all tongue-shaped.

💬 Experiment report

1. Write the main morphological characteristics of given familiars and list common pharmaceutical plants.

2. Write the plant flower formula for observed plants.

3. Draw the anatomic structure of a tubular bisexual flower, a tongue-shaped bisexual flower, a tongue-shaped female flower, and indicate the names of each part.

Questions

1. What the source of Chinese medicine 'Jinyinhua', and how to distinguish between the true and false?

2. What is the basis for the division of Asteraceae into two subfamilies?

3. What are the characteristics of Asteraceae plants that can explain their progressiveness, and their significance in adapting to the environment?

实验十二 禾本科、天南星科、百合科、姜科、兰科

目的要求

1. **掌握** 禾本科、天南星科、百合科、姜科和兰科药用植物的主要形态特征。

2. **熟悉** 对植物各器官特征的分类学描述方法，熟悉花的解剖方法并能写出花程式，学会使用植物分类检索表。

3. **识别** 禾本科、天南星科、百合科、姜科、兰科的主要药用植物。

实验材料

（一）禾本科

1. **代表植物** 薏苡带花全草及果实。

2. **其他标本** 芦苇带花全草。

（二）天南星科

1. **代表植物** 天南星带花序植株、果序。

2. **其他药用植物** 石菖蒲带花植株。

（三）百合科

1. **代表植物** 百合带花植株、鳞茎。

2. **其他药用植物** 黄精带花植株。

（四）姜科

1. **代表植物** 姜带花或果实的植株。

2. **其他药用植物** 姜黄带花植株。

（五）兰科

1. **代表植物** 天麻植物标本及新鲜植株。

2. **其他药用植物** 石斛带花植株；白及带花植株。

仪器用品

放大镜、解剖镜、解剖用具。

内容与方法

（一）观察解剖下列各代表植物

1. 禾本科

薏苡 取带果实的薏苡植株。观察植株、叶片、花序和果实。注意茎、叶的形态，花序、果实的形态和类型。

2. 天南星科

天南星　取带花序天南星植株。观察叶、佛焰苞、花序的形状和类型。注意叶的分裂、佛焰苞的颜色、附属体的形状、果实的类型。

3. 百合科

百合　取带花的百合植株。观察植株、鳞茎、花的形态、类型和颜色。注意鳞叶形态，花被片基部具蜜槽、花药的类型和颜色。取子房横切，观察子房室数及胎座类型。

4. 姜科

姜　取带花的姜，观察植株、根状茎、叶、花序、花的形态和类型。注意根状茎折断后芳香及辛辣味，叶片具有抱茎的叶鞘。注意花的着生位置、花瓣的形状和颜色，药隔具长喙状附属体。

5. 兰科

天麻　取带块茎和花的天麻植株。观察植株、块茎、叶和花的形态和类型。注意块茎具明显环节，叶呈膜质鳞片状，总状花序顶生，花冠下部壶状。

（二）观察其他药用植物标本

1. 禾本科

芦苇　取带根状茎和花的芦苇植株。观察植株、根状茎、叶片、花序的形态和类型。注意根状茎粗壮，叶舌有毛，花序分枝纤细并呈扫帚状。

2. 天南星科

石菖蒲　取带根状茎和花序的石菖蒲植株。注意植物气味、叶片、佛焰苞的形状和颜色，佛焰苞不包被花序。

3. 百合科

黄精　取带根状茎和花的黄精植株。观察植物、根状茎、叶、花的形态和类型。注意根状茎较长，叶先端卷曲，花裂片先端具乳突。子房横切，注意子房位置，子房室数和胎座类型。

4. 姜科

姜黄　取带根状茎和花的姜黄植株。观察植物的气味，块根、叶和花的形态和类型。注意茎横断面和根茎的颜色。注意花萼一侧开裂。注意花冠和退化雄蕊的颜色，花药基部两侧有距。

5. 兰科

（1）石斛　取带花的石斛植株。观察茎、叶、花的形态和类型。注意茎节多而明显。注意花、合蕊柱的形态和颜色。唇瓣近基部中央有一个深紫色大斑块。

（2）白及　取带花的白及植株。观察植物、花的形态和类型。注意花的对称性，花被片的排列和大小。注意子房位置，胎座类型，合蕊柱和花粉块的形态。

💬 **实验报告**

1. 阐述上述五个科的主要特征。
2. 写出所观察植物的花程式。
3. 编写上述植物的检索表。
4. 使用植物分类检索表，选择一种植物写出其检索过程。

📔 **思考题**

1. 生活中禾本科植物的主要用途有哪些？
2. 假种皮和种皮有什么不同？

Experiment 12 | Gramineae, Araceae, Liliaceae, Zingiberaceae, Orchidaceae

Aim and demand

1. Grasp the main morphological characteristics of Gramineae, Araceae, Liliaceae, Zingiberaceae and Orchidaceae.

2. Familiarize with the description of plant morphology and anatomical method. Familiarize with the anatomy of flower and write the floral formula. Learn how to use the Plant Classification Key to identify plants.

3. Recognize the main medicinal plants of the above families.

Experiment materials

1. Gramineae

1.1 Typical plant　The herb with flowers and fruits of Yiyi (*Coix lacryma-jobi* L. var. *ma-yuen* (Roman.) Stapf.).

1.2 Other medicinal plants　The plant with flowers of Luwei (*Phragmites communis* Trin.)

2. Araceae

2.1 Typical plant　The plant with inflorescence and infructescence of Tiannanxing (*Arisaema consanguineum* Schott.).

2.2 Other medicinal plants　The plant with flowers of Shichangpu (*Acorus tatarinowii* Schott.)

3. Liliaceae

3.1 Typical plant　The plant with flowers and bulb of Baihe (*Lilium brownii* F. E. Brown. var. *viridulum* Baker).

3.2 Other medicinal plants　The plant of Huangjing (*Polygonatum sibiricum* Redoute.) with flower.

4. Zingiberaceae

4.1 Typical plant　The plant with flowers and fruits of Jiang (*Zingiber officinale* Rosc.)

4.2 Other medicinal plants　The plant with flowers of Jianghuang (*Curcuma longa* L.).

5. Orchidaceae

5.1 Typical plant　Herbarium or fresh plant of Tianma (*Gastrodia elata* Bl.).

5.2 Other medicinal plants　The plant with flowers of Shihu (*Dendrobium nobile* Lindl.); and Baiji (*Bletilla striata* (Thunb.) Reichb. f.).

Instruments and appliances

Magnifier, anatomical lens, dissector.

Contents and Procedures

1. Observe and dissect the typical plants

1.1 Gramineae

Coix lacryma-jobi var. *ma-yuen*　Take the plant of *Coix lacryma-jobi* var. *ma-yuen* with flowers and fruits. Observe the shape and types of plant, leaves, inflorescence and fruits. Note the structure of leaves, inflorescence and fruit.

1.2 Araceae

Arisaema consanguineum　Take the plant of *Arisaema consanguineum* with the inflorescence. Observe the shape and type of leaf, spathe, inflorescence. Note the leaf lobes, the color of spathe, the shape of appendages, and the type of fruit.

1.3 Liliaceae

Lilium brownii var. *viridulum*　Take the plant of *Lilium brownii* var. *viridulum* with flowers. Observe the shape and color of herb and flower. Pay attention to the shape of leaf, the base of tepals with honey groove, the type and color of anther. Observe the transection of ovary and the number of the locule and type of placenta.

1.4 Zingiberaceae

Zingiber officinales　Take the plant of *Zingiber officinales* with flowers. Observe the shape and type of plant, rhizome, leaves, inflorescence and flower. Note the aromatic and pungent taste of the broken rhizome, and the leaves have a sheath that hugs the stem. Note the position of flowers, the shape and color of petals, and the connective with a long beaked appendages.

1.5 Orchidaceae

Gastrodia elata　Take the plant of *Gastrodia elata* with flower. Observe the shape and type of plant, herb, leaf and flower. Note tuber with distinct segments, leaves membranous and scaly, racemes terminal, lower corolla ampulla.

2. Observe other samples of medical plants

2.1 Gramineae

Phragmites communis　Take the plant of *Phragmites communis* with rhizome. Observe the shape and type of plant, rhizome, leaf and inflorescence. Note the stout rhizome, hairy leaf tongue, inflorescence branches slender and broomlike.

2.2 Araceae

Acorus tatarinowii　Take the plant of *Acorus tatarinowii* with rhizome and inflorescence. Pay attention to the odour of plant, the shape and color of leaf and spathe. Note the spathe is not enveloped by inflorescence.

2.3 Liliaceae

Polygonatum sibiricum　Take the plant of *Polygonatum sibiricum* with rhizomes and flowers. Observe the shape and type of plant, rhizome, leaf and flower. Note the rhizome is long, leaf apex curly, flower lobes apex papillate. Transversely cut ovary and observe the number of ovary locule and type of placenta.

2.4 Zingiberaceae

Curcuma longa　Take the plant of *Curcuma longa* with rhizome and flower. Observe the odour of plant, the shape and type of tuber, leaf and flower. Note the color of the stem cross section and rhizome.

Note that one side of the calyx is dehiscent. Pay attention to the color of corolla and staminodes, with spaces on both sides of anthers base.

2.5 Orchidaceae

2.5.1 *Dendrobium nobile*   Take the plant of *Dendrobium nobile* with flowers. Observe the shape and type of stem, leaves and flowers. Note that the stem nodes are numerous and conspicuous. Note the shape and color of the flower and stamen column. Pay attention to a large dark purple patch in the center of the lower lobe near the base.

2.5.2 *Bletilla striata*   Take the plant of *Bletilla striata* with flowers. Observe the shape and type of plant and flower. Note the symmetry of the flower, the arrangement and size of perianth. Note ovary location, placental type, stamen column and pollen block morphology.

Experiment report

1. What are the main characteristics of the given families.

2. Write the plant flower formula for observed plants.

3. Compile the Plant Classification Key for observed plants.

4. Using the Plant Classification Key during the observation, choose a plant and describe the identification process.

Questions

1. What are the main uses of Gramineae plants in our life?

2. What's the difference between aril and a seed coat?

实验十三 植物细胞的结构和后含物

目的要求

掌握 植物细胞的基本构造；植物细胞后含物的显微特征。

实验材料

洋葱鳞叶、马铃薯块茎、半夏粉末、黄柏粉末、大黄粉末、印度橡皮树叶横切片、桔梗茎横切片。

仪器用品

光学显微镜、镊子、解剖针、刀片、载玻片、盖玻片、培养皿、吸水纸、擦镜纸、酒精灯、蒸馏水、稀碘液、水合氯醛、稀甘油。

内容与方法

（一）植物细胞的基本构造

1. **制片** 取洋葱鳞叶 1 片，在其内表面用锋利刀片划出纵横的平行线若干，形成 3~4mm 见方的小方格。用镊子仔细揭取 1 小片表皮。在载玻片中央滴 1 滴蒸馏水，将表皮置于水滴中。然后用镊子夹住盖玻片一边，使其另一边接触水滴，慢慢放下盖玻片，以避免气泡的产生。如盖玻片下的水过多，可用吸水纸从一侧吸去多余的水。即制成水装片。

2. **观察** 将制好的洋葱鳞叶表皮水装片置于低倍镜下观察，可看到许多无色透明的、稍长形的细胞，彼此紧密相连，没有细胞间隙。移动标本片，选择数个较清晰的细胞于视野中央，转换高倍镜并调节焦距至清晰，可见有以下结构。

（1）细胞壁 位于原生质体的最外面，由于细胞壁是无色透明的结构，所以只能看到每个细胞四周的侧壁。

（2）细胞质 在细胞壁的内侧，有一圈半透明的薄层，即为细胞质膜。表皮细胞是成熟细胞，由于中央液泡的形成，细胞质被挤压到贴近细胞壁处。

（3）细胞核 位于细胞的中央或靠近细胞壁的细胞质中，在细胞中央的多为圆球形，靠近细胞壁的多为扁球形或半圆形。如加稀碘液染色，细胞核被染成黄褐色。转动细准焦螺旋，可见在细胞核中有一至数个较亮的小球形体，即核仁。

（4）液泡 位于细胞中央，比细胞质更为透明，其内充满细胞液。如加稀碘液，液泡不被染色，仍是透明无色的。

（二）植物细胞的后含物

1. **淀粉粒** 取 1 小块马铃薯块茎，用刀片轻轻刮取少量液汁，置于载玻片上，加水制成水装片观察。在低倍镜下可见到许多类圆形的颗粒，即淀粉粒。转换高倍镜，仔细观察脐点和层纹，注意分辨单粒、复粒和半复粒，绘图记录。然后取下制片，加1滴稀碘液，观察有何变化。

2. 草酸钙结晶

（1）针晶　取半夏粉末制成水合氯醛透化片，可见散在或成束的针状草酸钙晶体。

（2）方晶　取黄柏粉末制成水合氯醛透化片，可见方晶整齐排列于纤维束周围，这种复合体称为晶鞘纤维。

（3）簇晶　取大黄粉末制成水合氯醛透化片，可见由不规则三棱形或多面体，单晶体聚集而成的簇晶。

3. 碳酸钙结晶　观察印度橡皮树叶的横切片，在叶的表皮细胞的大型细胞中有一个葡萄状的结晶体附着在细胞壁增生的棒状物上，悬挂在细胞腔中，形似钟乳体，即为碳酸钙晶体。

实验报告

1. 绘制洋葱鳞叶内表皮细胞构造图，并注明细胞各主要部分。
2. 绘制马铃薯淀粉粒形态图。
3. 绘制三种草酸钙结晶的形态图。

思考题

如何区分草酸钙结晶和碳酸钙结晶？

习题

Experiment 13　Structure and Main Ergastic Substance of Plant Cells

Aim and demand

Grasp the basic structure of plant cells and the microscopic characteristics of ergastic substances of plant cells.

Experiment materials

Bulb of Yangcong (*Allium cepa* L.), tuber of Malingshu (*Solanum tuberosum* L.), Banxia (Pinella ternata (Tihunb.) Breit), powder of Huangbai (phellodendron amurense Rupr.), powder of Dahuang (Rheumpalouatum L.), slide of cross section of leaf of Yinduxiangpishu *Ficus elastica* Roxb ex Hornem, slide of cross section of Jiegeng *Platycodon grandiflorum* (Jacq.) A. DC.

Instrument and appliances

Microscope, tweezers, dissecting needle, scalpel, slide, cover glasses, culture dish, blotting paper, lens paper, alcohol lamp, distilled water, dilute iodine, chloral hydrate, dilute glycerin.

Contents and Procedures

1. The structure of plant cells

1.1　Samples　Take a scale leaf of onion and carve several parallel lines of 3-4 mm with a scalpel on its inner surface to form a small square. Remove one small piece of cuticle with tweezers. Place it in the drop of distilled water in the center of the slide. Then hold one side of the cover glass with tweezers so that the other side is in contact with the water drop. Slowly lower the cover glass to avoid bubbles. If the water under the cover glass is too much, use blotting paper from one side to absorb the excessive water. A temporary mount microscope slide is made.

1.2　Observation　Many colorless and transparent slightly long cells can be found under the low power microscope. The cells are closely connected with each other. Select several clearer cells in the center of the vision, then switch to the high power microscope. Adjust the focal length until the following structure can be seen clearly.

1.2.1　Cell wall　The cell wall is the outermost part of the protoplast. Only the side walls around each cell can be seen because it is colorless and transparent.

1.2.2　Cytoplasm　Inside the cell wall, there is a translucent ring of thin layer called cytoplasmic membrane. Epidermal cells are mature cells whose cytoplasm is squeezed closely to the cell wall by the formation of central vacuole.

1.2.3 Nucleus   The nucleus is in the center of the cell or near the cell wall. If the cell is stained with dilute iodine, the nucleus is stained yellowish-brown. Turn the fine adjustment, one or several globular and lighter globular bodies, the nucleoli are visible in the nucleus.

1.2.4 Vacuoles   Vacuoles are more transparent than cytoplasm and are filled with cell fluid. Add dilute iodide solution, vacuole is still transparent colorless.

2. Starch grain   Take one small tuber of *Solanum tuberosum* L., gently scrape a little of liquid with a scalpel, put it in the water which was already on a glass slide. Many round granules can be seen at low power microscope, then turned to high power microscope. Umbilicus and annular striation lamellae can be seen. Pay attention to the simple starch grain, compound and half-compound starch grain. Then add one drop of dilute iodine to see what happens.

3. Calcium oxalate crystal

3.1 Needle crystal   The permeable tablets of chloral hydrate were prepared from the powder of pinellaternata, scattered or bunched needle crystals can be seen.

3.2 Solitary crystal   The permeable tablets of chloral hydrate were prepared from the powder of phellodendron amurense, you can see that the solitary crystals are arranged in order around the fiber bundles, which are called crystal fiber.

3.3 Cluster crystal   The permeable tablets of chloral hydrate were prepared from the powder of Rheumpalouatum, you can see clusters which are formed by many irregular triangular or polyhedral single crystals.

4. Calcium carbonate crystal   Observe the transverse section of leaf of *Ficus elastica*. In the large cells of epidermis, there is a grape like crystal attached to the proliferative rod of cell wall, hanging in the cell cavity, which looks like a bell emulsion, namely calcium carbonate crystal.

Experiment report and questions

1. Draw the structure figure of epidermis cells of onion scale leaf, and note the name of every section.

2. Draw the structure chart of potato starch grains.

3. Draw the morphology of three kinds of calcium oxalate crystals.

Questions

How to distinguish calcium oxalate crystal from calcium carbonate crystal?

实验十四　植物组织和维管束

目的要求

1. **掌握**　植物六大组织的形态特征及类型。
2. **熟悉**　各种组织在植物体内的主要分布位置。
3. **了解**　维管束的基本构造，识别维管束的类型。

实验材料

1. **新鲜材料**　薄荷叶、菘蓝叶、曼陀罗叶、菊叶、毛茛叶、茶叶、洋地黄叶、蒲公英根、姜根茎。
2. **药材**　肉桂粉末、厚朴粉末、苦杏仁。
3. **石蜡切片**　洋葱根尖纵切面、薄荷茎横切面、接骨木幼茎横切面、厚朴树皮横切面、陈皮横切面、松茎纵切面、松茎横切面、南瓜茎纵切面、向日葵茎横切面、玉米茎横切面、粗茎鳞毛蕨根茎横切面、石菖蒲根茎横切面、毛茛根横切面。

仪器用品

显微镜、解剖工具、稀甘油、水合氯醛试剂、蒸馏水、间苯三酚、浓盐酸、苏丹Ⅲ试液、20% 乙酸溶液。

内容与方法

（一）分生组织

1. **初生分生组织**　取洋葱根尖纵切面制片，置于低倍镜下观察，可见在根尖的先端有许多排列疏松的细胞群组成一个帽状的根冠，内侧就是根的初生分生组织——生长点。注意根尖生长点细胞的细胞壁、细胞质和细胞核的特征以及细胞排列情况。

2. **次生分生组织**

（1）维管形成层　取薄荷茎横切片，置于显微镜下观察，可见排列成环状的维管束。在维管束的木质部（细胞被染成红色部位）与韧皮部（染成深绿色的一群细胞）之间，可见排列整齐、紧密略呈扁长方形的细胞群，这就是维管形成层。

（2）木栓形成层　取接骨木幼茎横切片，置显微镜下观察，可见规则排列的几层木栓细胞。木栓细胞的特点是扁平形、排列整齐、褐色且无细胞间隙。在木栓细胞内有 1~3 层含有原生质体，并处于分裂状态的活细胞，为木栓形成层。木栓细胞形成层切向（即平行于表皮的）横壁分裂，往外产生木栓化的、死亡细胞（木栓层），而向内则产生活的薄壁细胞（栓内层）。由木栓层、木栓形成层和栓内层细胞形成次生性复合保护组织——周皮。

（二）保护组织

1. **表皮与毛茸**　用镊子撕取薄荷叶下表皮，制作临时装片，于显微镜下观察，注意表皮细胞的形状和排列特点。在表皮细胞上长有多细胞的先端尖锐的毛茸为非腺毛。此外，有些毛茸有类圆球状的头部与短柱状柄部之分的为腺毛。注意薄荷叶腺毛的头部和柄部各由多少细胞组成。

2. **气孔**

（1）平轴式气孔　撕取薄荷叶下表皮，制作临时装片，于显微镜下观察，可见气孔周围有2个副卫细胞，其长轴与保卫细胞和气孔的长轴平行。

（2）直轴式气孔　撕取薄荷叶下表皮，制作临时装片，于显微镜下观察，可见气孔周围2个副卫细胞，其长轴与保卫细胞和气孔的长轴相垂直。

（3）不等式气孔　撕取菘蓝叶或曼陀罗叶的下表皮，制作临时装片，于显微镜下观察，可见气孔保卫细胞周围有3个副卫细胞，其中1个显著地比其他副卫细胞小。

（4）不定式气孔　撕取菊或毛茛的下表皮，制作临时装片，于显微镜下观察，可见副卫细胞数目不定，其形状与其他表皮细胞基本相似。

（5）环式气孔　撕取茶叶表皮，制作临时装片，于显微镜下观察，可见气孔周围的副卫细胞数目不定，比表皮细胞狭窄，排列成环状。

（三）薄壁组织

观察薄荷茎横切片，注意位于厚角组织两侧以及茎的最中央有许多类圆形的薄壁细胞，细胞与细胞之间有明显的间隙，该细胞群即为薄壁组织。

（四）机械组织

1. **厚角组织**　观察薄荷茎横切片，可见茎棱角处的表皮下方，有数层细胞，其细胞只在角隅处增厚，即为厚角组织。

2. **厚壁组织**　即纤维、石细胞。取肉桂粉末少许，以水合氯醛装片，置于显微镜下观察，可见某些细胞呈长梭形，细胞壁很厚，胞腔很小或看不清胞腔，这种细胞为纤维。

用镊子撕取苦杏仁种皮适量，以水合氯醛装片，进行观察，可见多数散在的黄色类圆形石细胞，其细胞壁增厚，并有许多纹孔。

（五）输导组织

1. **管胞**　取松茎纵切片，于显微镜下观察，可见众多两端斜尖的长梭形管状细胞。相邻两管壁上的纹孔相同，为具缘纹孔，调节细准焦螺旋，可见三个同心圆。

2. **导管及筛管**　观察南瓜茎纵切片，可见被番红染料染成红色的长管状组织，即为导管群，注意观察南瓜茎有几种纹理的导管，每个导管细胞均以端壁形成的穿孔相互连接。导管群两侧染成绿色的薄壁性纵行连接的管状组织，即为筛管群。注意观察上下两个筛管细胞相接的横隔特征，该横隔就是筛板。

（六）分泌组织

1. **分泌细胞**　制作生姜临时装片，可见椭圆形的细胞，充满黄色溶液，该细胞即为油细胞，黄色溶液即为挥发油。

2. **分泌腔**　观察陈皮横切片，可见由许多薄壁细胞围拢而成的大型圆形腔隙，即为分泌腔。陈皮的分泌腔内含有挥发油，又称为油室。

3. **分泌道**　观察松茎横切片，可见由许多分泌细胞围成的圆形的腔隙，即分泌道，因其内含有树脂，又称树脂道。

4. **乳汁管**　制作蒲公英根纵向临时切片，在显微镜下可见颜色略深的分枝状乳汁管。

（七）**维管束的构造和类型**

取向日葵茎横切片，先观察韧皮部——筛管及伴胞，木质部——导管，注意二者的排列方式；然后观察木质部和韧皮部之间细胞的特征，指出维管束类型。

同样方法观察下列永久切片：①玉米茎横切片——有限外韧型维管束；②南瓜茎横切片——双韧型维管束；③粗茎鳞毛蕨根茎横切片——周韧型维管束；④石菖蒲根茎横切片——周木型维管束；⑤毛茛根横切片——辐射型维管束。

💬 实验报告

1. 简述各类组织的形态特征及类型。
2. 绘出薄荷叶、洋地黄叶的气孔器，注明各部分的名称。
3. 绘出肉桂的纤维、苦杏仁的石细胞，注明各部分的名称。
4. 绘出生姜的分泌细胞、陈皮的油室及蒲公英的乳汁管。
5. 绘各种维管束类型简图，注明各部位名称及材料名称。

🖊 思考题

1. 分生组织的细胞形态特点是什么？
2. 气孔的作用是什么？具缘纹孔是如何形成的？
3. 维管束一般构造如何？可分为几种类型？

习题

Experiment 14 　Plant Tissues and Vascular Bundle

📋 Aim and demand

1. Grasp the morphology, type and structure of six kinds of plant tissues.
2. Familiarize the distribution of various plant tissues.
3. Understand the basic structure of vascular bundle, distinguish the types of vascular bundle.

⚫ Experiment materials

1. Fresh materials The leaves of Bohe (*Mentha haplocalyx* Briq.), Songlan (*Isatis indigotica* Fort.), Mantuoluo (*Datura stramonium* L.), Ju (*Chrysanthemum morifolium* Ramat.), Maogen (*Ranunculus japonicus* Thunb.), Cha (*Camellia sinensis* (L.) O. Ktze.), Yangdihuang (*Digitalis purpurea* L.), root of Pugongying (*Taraxacum mongolicum* Hand-Mazz) and rhizome of Jiang (*Zingiber officinale* Rosc.).

2. Crude drugs The powder of Rougui (Cinnamomi Cortex), Houpo (Magnoliae Officinalis Cortex), Kuxingreng (Armeniacae Semen Amarum.)

3. Paraffin sections Longitudinal section of the root tip of Yangcong (*Allium cepa* L.), cross section of the stem of Bohe (*Mentha haplocalyx* Briq.), young stem of Jiegumu (*Sanbucus williamsii* Hance), Houpo (Magnoliae Officinalis Cortex), Citri Reticulatae Pericarpium, the longitudinal section of the stem of Song (*Pinus massoniana* Lamb.) and Chenpi (*Cucurbita moschata* (Duch. ex Lam.) Duch. ex Poiret), cross section of the stem of Song (*Pinus massoniana* Lamb.), Xiangrikui (*Helianthus annuus* L.) and Yumi (*Zea mays* L.), the rhizome of Jue (*Dryopteris crassirhizoma* Nakai), Shicangpu (*Acorus tatarinowii* Schott) and the root of Maogen (*Ranunculus japonicus* Thunb.).

⚙ Instrument and appliances

Microscope, dissection tools, dilute glycerine, chloral hydrate, distilled water, phloroglucinol, concentrated hydrochloric acid, Sudan Ⅲ solution, 20% acetic acid solution.

🔍 Contents and Procedures

1. Meristem

1.1 Primary meristem Take a longitudinal section of the root tip of *Allium cepa* L., and observe under lower power lens. There are many loose cell groups arranged at the apex of the root tip to form calyptra, and the inner side is the primary meristem of the root, which is growing point. Observe the characteristics and cell arrangement of cell wall, cytoplasm, nucleus of the root tip of growing point.

1.2 Secondary meristem

1.2.1 Vascular cambium　Take a cross section of the stem of *Mentha haplocalyx* Briq., and observe under microscope. Circular vascular bundle can be seen. Between the xylem (where the cells are dyed red) and the phloem (a group of cells dyed dark green), the orderly arranged, compact and slightly oblong cell groups is the vascular cambium.

1.2.2 Phellogen   Take a cross section of young stem of *Sanbucus williamsii* and observe it under microscope. Several layers of brown flat cork cell arranged in order, without cell interspace can be seen. Beneath the cork cells, 1-3 layers of living cells contained protoplast, called phellogen are in division state. Phellogen cells split along tangent walls (parallel to epidermis), which generate suberized dead cells outside, and nonsuberized cells with live protoplast (phelloderm cells) inside. Periderm is the complicated protective tissue composed by suberized cork cells, phellogen and phelloderm cells.

2. Protective tissue

2.1 Epidermis and epidermal hair   Tear off the lower epidermis of the leaves of *Mentha haplocalyx* and make a temporary water slice and observe under microscope. Pay attention to the shapes and arrangement characteristics of epidermal cells. Epidermal hair with sharp apex on epidermal cells, which is called non-glandular hair. In addition, some epidermal hair, whose apex are not sharp but with spherical head and stub like petiole, are called the glandular hair. Pay attention to the number of head or stub cells of epidermal hair.

2.2 Stoma

2.2.1 Paracytic type   Tear off the lower epidermis of the leaves of *Mentha haplocalyx*, and make a temporary water slice and observe under microscope. Two subsidiary cells should be visible around the stoma, and its long axis is parallel to the long axis of guard cell and stoma.

2.2.2 Diacytic type   Tear off the lower epidermis of the leaves of *Mentha haplocalyx*, make a temporary water slice and observe under microscope. Two subsidiary cells should be visible around the stoma, and its long axis is vertical to the long axis of guard cell and stoma.

2.2.3 Anisocytic type   Tear off the lower epidermis of the leaves of *Isatis indigotica* or *Datura stramonium*, make a temporary water slice and observe under microscope. Three subsidiary cells should be visible around the stoma, one of which appears smaller than two others.

2.2.4 Anomocytic type   Tear off the lower epidermis of the leaves of *Chrysanthemum morifolium* or *Ranunculus japonicus*, make a temporary water slice and observe under microscope. A variable number of subsidiary cells should be visible around the stoma, and its shape is similar to the common epidermis cells.

2.2.5 Actinocytic type   Tear off the epidermis of the leaves of *Camellia sinensis*, make a temporary water slice and observe under microscope. A variable number of subsidiary cells should be visible around the stoma, which is narrower than the common epidermis cells, and arrange in a ring.

3. Parenchyma

Take a cross section of the stem of *Mentha haplocalyx* and observe it under microscope. Pay attention that many subrotund parenchyma cells locat on both sides of the collenchyma, as well as in the center of the stem, and the interspace between these cells is obvious.

4. Mechanical tissue

4.1 Collenchyma   Observe a cross section of the stem of *Mentha haplocalyx* under microscope. As a result, several layers of cells under the epidermis at the corner of the stem. These cells only thicken at the corner, which is called collenchyma.

4.2 Sclerenchyma   It is fiber and stone cell. Take a bit of powder of Cinnamomi Cortex, mount with chloral hydrate and test under microscope. We can see some fusiform cells with thick wall and small or unclear cell lumen, which are called fiber.

Tear off testa of Armeniacae Semen Amarum, mount with chloral hydrate and test under microscope. We can see the scattered yellow subrotund stone cells with thick walls and many pits.

5. Conducting tissue

5.1 Tracheid　Observe a longitudinal section of the stem of *Pinus massoniana* under microscope, many fusiform tubiform cells with cuspidal end called tracheid can be seen. The pits on the two adjacent cell walls are the same, which are the bordered pit. Furthermore, three concentric circles can be found by regulating the fine adjusting screw.

5.2 Vessel and sieve tube　Test a longitudinal section of the stem of *Cucurbita moschata* under microscope and the red tubiform tissue dyed by safranine called vessel mass that can be found. Pay attention to the types of vessel. Vessel cells are connected with each other by the perforation formed by the end wall. The tubiform tissue dyed green on both sides of vessel mass is longitudinal-connected sieve tubes with thin walls and the septum between two longitudinal-connected sieve tubes, which is sieve plate.

6. Secretory tissue

6.1 Secretory cell　Make temporary slice of the rhizome of *Zingiber officinale*, and test under microscope. Elliptic cells that are full of yellow solution, the volatile oil, are called the secretory cell.

6.2 Secretory cavity　Test a cross section of Citri Reticulatae Pericarpium under microscope and the large round lacounas surrounded by many parenchyma cells called secretory cavity can be found. Due to volatile oil contained in secretory cavity, therefore, it is also called oil cavity.

6.3 Secretory canal　Test a cross section of the stem of *Pinus massoniana* under microscope. And we can see many rounded lacounas formed by many secretory cells, which is secretory canal. The lacounas of secretory canals contain resin in pine stem, so they are called resin canal.

6.4 Laticifer　Make the temporary longitudinal section of root of *Taraxacum mongolicum* Under the microscope, the slightly dark branching laticifer can be seen.

7. Structure and types of vascular bundle

Take cross section of the stem of *Helianthus annuus*. Firstly, to observe the phloem-sieve tube and companion cell, xylem-vessel, pay attention to the order of the phloem and xylem. Then observe the cell characteristics of the phloem and xylem, point out the type of vascular bundle.

Observe slices below use the same method as above.

Cross section of the stem of *Zea mays*-closed collateral vascular bundle.

Cross section of the stem of *Cucurbita moschata*-bicollateral vascular bundle.

Cross section of the rhizome of *Dryopteris crassirhizoma*-amphicribrial vascular bundle.

Cross section of the rhizome of *Acorus tatarinowii*-amphivasal vascular bundle.

Cross section of the root of *Ranunculus japonicus*-radial vascular bundle.

Experiment report

1. State the morphologies, types and structures of six kinds of plant tissues.

2. Draw the stoma of the leaves of *Mentha haplocalyx*, *Digitalis purpurea* and indicate the names of each part.

3. Draw fibers of the powder of Cinnamomi Cortex, stone cells of Armeniacae Semen Amarum and indicate the names of each part.

4. Draw secretory cell of rhizome of *Zingiber officinale*, oil cavity of Citri Reticulatae Pericarpium and laticifer of root of *Taraxacum mongolicum*.

5. Draw sketch of all types of vascular bundles you have observed, note names of different parts and materials.

Questions

1. What are the morphologic characteristics of the meristematic cells?
2. What are the functions of stoma? How the bordered pits are formed?
3. What is the general structure of vascular bundle? How many types of vascular bundles are there?

实验十五　根的构造

PPT

目的要求

1. **掌握**　双子叶植物根和单子叶植物根的初生构造及异同点。
2. **掌握**　双子叶植物根的次生构造特征。
3. **了解**　双子叶植物根的异常构造。

实验材料

毛茛、辽细辛、鸢尾、百部、芍药、人参、怀牛膝、何首乌、黄芩根的永久横切片。

仪器用品

显微镜。

内容与方法

（一）双子叶植物根的初生构造

取毛茛根的横切片，在低倍镜下区分出表皮、皮层和维管柱三大部分，然后转换高倍镜由外向内仔细观察。

1. 表皮　为根最外面的一层薄壁细胞，排列紧密整齐，没有细胞间隙，注意观察表皮上有无角质层和根毛。

2. 皮层　在表皮以内，占根体积的绝大部分，由多层排列疏松的薄壁细胞组成，可分为如下三部分。

（1）外皮层　紧靠表皮下方，排列较紧密，略呈切向延长的1~2层薄壁细胞。

（2）皮层薄壁组织　介于外皮层和内皮层之间，由排列疏松的多层大型薄壁细胞组成。

（3）内皮层　为皮层最内一层排列紧密的细胞组成，细胞径向壁上可见被染成红色的凯氏点，对着木质部束有细胞壁不增厚的通道细胞。

3. 维管柱　为内皮层以内的所有组织，占据根中央小部分，细胞小而密集，由如下三部分组成。

（1）中柱鞘　由紧贴内皮层的1~2层小型薄壁细胞组成，排列整齐而紧密。中柱鞘细胞具有潜在的分生能力。

（2）初生木质部　呈四束（四原型），导管被染成红色，导管直径大小不一，靠近中柱鞘的导管直径小，为原生木质部；近根中心的导管直径较大，为后生木质部。这是根的初生构造特征之一。

（3）初生韧皮部　位于相邻两个初生木质部之间，与初生木质部相间排列，构成辐射型维管束，这也是根的初生构造特征之一。初生韧皮部被染成绿色，细胞大小不一，形态不规则。

此外，在初生木质部和初生韧皮部之间分布着薄壁细胞，当根进行次生生长时，参与分化形成维管形成层。

同样观察细辛根的横切片，注意区分表皮、皮层和维管柱。其初生木质部通常三原型。

（二）单子叶植物根的构造

观察鸢尾根的横切片，注意区分表皮、皮层、维管柱和髓四部分。

1. 表皮 为根最外面一层细胞，有的残存。

2. 皮层 外皮层细胞稍木栓化，增强保护作用。内皮层细胞的多数细胞除外切向壁的其他五面全加厚，并且木栓化和木质化，横切面呈马蹄形。正对原生木质部尖端的薄壁细胞被称为通道细胞。

3. 维管柱 为内皮层以内的所有组织，占据根中央小部分，细胞较小而密集，由三个部分组成。

（1）中柱鞘 紧贴内皮层，1~2 层小型薄壁细胞，细胞切向延长略似内皮层细胞，但侧壁没有增厚。

（2）初生木质部和初生韧皮部 多原型，19~27 个，相间排列成辐射型维管束，二者之间有少量薄壁组织。

4. 髓 位于维管柱的中心，髓细胞在后期可能转变成木质化的厚壁细胞。

同样方法观察百部根的横切片，注意区分其各组成部分。

（三）双子叶植物根的次生构造

取芍药根的横切片，从外向内观察。

1. 周皮 为最外方的数层细胞，由木栓层、木栓形成层和栓内层组成。

（1）木栓层 由多层排列整齐、扁长形的木栓细胞组成，常呈浅棕色。

（2）木栓形成层 由中柱鞘细胞恢复分生能力产生，在切片中不易分辨。

（3）栓内层 为 2~3 层呈切向延长的薄壁细胞。

2. 次生维管组织 为形成层活动产生的组织。

（1）次生韧皮部 周皮以内被染成绿色的部分，较宽。包括筛管、伴胞、韧皮薄壁细胞和少量韧皮纤维。在横切面上，韧皮薄壁细胞与筛管形态相似，常不易区分。韧皮射线呈放射状，由韧皮部薄壁细胞在径向上排列而成。

（2）维管形成层 在次生韧皮部的内方。形成层只有一层细胞，但由于其切向分裂迅速，刚产生不久的衍生细胞尚未分化成熟，所以在横切面上看到的是由多层排列整齐、紧密的扁长形薄壁细胞组成的"形成层区"。

（3）次生木质部 在形成层以内，包括导管、管胞、木薄壁细胞和木纤维。在横切面上，导管最易辨认，是被染成红色的、直径大小不一的一些类圆形或多边形的死细胞，呈放射状排列。木射线由 1~2 列薄壁细胞组成，在木质部中呈放射状排列，并与韧皮射线相连接，组成维管射线，起横向运输的作用。

在次生木质部的内方，即根的中心部位，为初生木质部。其导管直径较小，呈类圆形。

同样方法观察人参根的横切片，其特点是次生韧皮部有树脂道；木质部导管多呈单列、径向稀疏排列；栓内层、木薄壁细胞和木射线中有草酸钙簇晶。

（四）根的异常构造

1. 观察何首乌块根药材，其横断面呈淡黄棕色或淡红棕色，皮部有 4~11 个异型维管束环列，形成"云锦花纹"。然后在显微镜下观察何首乌块根的横切片，从外向内区分出周皮、薄壁组织、异型维管束、正常维管束。形成层呈环状。异型维管束多为复合型，少数为单个维管束。根的中央为正常外韧型维管束，中心部分为初生木质部。

2. 观察怀牛膝根药材，可见其横断面平坦，木质部黄白色，其外有众多小点，排成2~4轮，即异型维管束。然后在显微镜下观察其横切片，外部为木栓层，由4~8层扁平的木栓化细胞组成，其内为数层薄壁细胞。维管组织占根的大部分，外部有2~4轮外韧型的异型维管束，根中央为正常维管束，初生木质部常二原型。

3. 观察黄芩老根的横切片，木质部位于根的中央，木质部中分布有木栓化细胞环。

实验报告

1. 绘制毛茛根、百部根的横切片简图，并注明各部分名称。
2. 绘制芍药根的横切片简图，注明各部分名称。
3. 绘制怀牛膝或何首乌根的横切片简图，注明各部分名称。

思考题

1. 双子叶植物和单子叶植物根的初生构造有什么异同点？
2. 双子叶植物根的初生构造和次生构造有何区别？次生构造是如何形成的？
3. 双子叶植物根的异常构造是如何形成的？

Experiment 15 | Structure of Root

Aim and demand

1. Grasp the primary structure and differences between the root of dicotyledon and monocotyledon.
2. Grasp the secondary structure of dicotyledon.
3. Understand the anomalous structure of dicotyledon root.

Experiment materials

Paraffin sections of the root of Maogen (*Ranunculus japonicus* Thunb.), Xixin (*Asarum heterotropoides* Fr. Schmidt var. mandshuricum (Maxim.) Kitag.), Yuanwei (*Iris tectorum* Maxim.), Baibu (*Stemona sessilifolia* (Miq.) Franch. et Sav.), Shaoyao (*Paeonia lactiflora* Pall.), Renseng (*Panax ginseng* C.A.Mey.), Huainiuxi (*Achyranthes bidentata* Bl.) and Huangqin (*Scutellaria baicalensis* Georgi), the tuber of Heshouwu (*Polygonum multiflorum* Thunb.).

Instrument and appliances

Microscope.

Contents and Procedures

1. Primary structure of dicotyledon root

Observe the cross section of root of *Ranunculus japonicus*, to distinguish epidermis, cortex and vascular cylinder under the low-power lens of microscope, then turn to the high-power lens to observe from outer to inner.

1.1 Epidermis It is a single layer of densely packed cells in the outer, without intercellular space. Paying attention to if there is cuticle or root hair.

1.2 Cortex It lies inside the epidermis, and occupies a much larger proportion of root. It is divided into three parts as follows.

1.2.1 Exodermis It lies closely under the epidermis, which is composed of 1—2 layers of closely arranged cells, with slightly tangential extension.

1.2.2 Cortex parenchyma It lies between the exodermis and the endodermis, and consists of loose multilayer of parenchyma cells.

1.2.3 Endodermis It's the innermost one layer of the cortex. Cells are arranged densely with red casparian spot in the radial wall, and passage cells at the opposited xylem bundles.

1.3 Vascular cylinder It lies inward the endodermis and occupies a small part of the root center. Cells are small and dense, and divided into three parts as follows.

1.3.1 Pericycle   It consists of 1—2 layers of closely and regularly arranged parenchyma cells. The pericycle cells are potential meristems.

1.3.2 Primary xylem   It is made up of 4 bundles, namely tetrach. Its vessel is red and the diameter of each vessel is different. The small diameter of vessel near pericycle is named protoxylem; the big one of vessel near center is named metaxylem. This is one of the primary structural features of roots.

1.3.3 Primary phloem   It lies between two primary xylem by alternate arrangement with primary xylem and compose to radial vascular bundle. This is also one character of the primary structure of roots. This area is stained green with irregular cells of varying sizes.

In addition, there is a thin layer of cells between primary xylem and primary phloem. When the root undergoes secondary growth, they will differentiate into a part of vascular cambium.

Observe cross section of root of *Asarum heterotropoides* using the same method, and pay attention to distinguishing epidermis, cortex and vascular cylinder. Primary xylem is usually triarch.

2. Primary structure of monocotyledon root.

Observe the cross section of root of *Iris tectorum*, and distinguish the epidermis, cortex, vascular column and pith.

2.1 Epidermis   It is the outermost layer of residual cells of the root.

2.2 Cortex   Some part is remnant, and cuticle cell under the epidermis are slightly corked, which enhanced the protective effect. Pay attention to the endodermis. Five sides except outer tangent wall of most cells are thickened and corked and lignified. The cross-section is horseshoe-shaped thickening. The parenchyma cell on the tip of the primary xylem is called passage cell.

2.3 Vascular cylinder   The whole tissue lies inward the endodermis, which is divided into three sections:

2.3.1 Pericycle   1—2 layers of small parenchyma cells close to the endodermis. The tangential extension of cells slightly like the endodermis, but the lateral wall is not thickened.

2.3.2 Primary xylem and primary phloem   Multiple prototypes, with 19—27 bundles in each, arranged as radiation vascular bundles. There are a little parenchyma between primary xylem and primary phloem.

2.4 Pith   Located in the center of vascular cylinder, pith cells may be transformed into lignified sclerenchyma cells in later growth stage.

Observe the cross section of root of *Stemona sessilifolia* in the same way, and distinguish the different parts.

3. The secondary structure of dicotyledon

Take cross section of root of *Paeonia lactiflora* to observe from outer to inner.

3.1 Periderm   It is composed of lots of layers of cells at the outer, including phellem, phellogen and phellodem.

3.1.1 Phellem   It consists of flat rectangular cork cells of regularly arranged, which are light brown.

3.1.2 Phellogen   It is produced by the recovery of mitogenetic ability of pericycle cells, which is difficult to distinguish in sections.

3.1.3 Phelloderm   It is 2—3 layers of parenchyma cells with tangential extension.

3.2 Secondary vascular tissue   The tissue resulted from vascular cambium.

3.2.1 Secondary phloem   It is dyed green and wide within the periderm, and includes sieve tube, companion cell, phloem parenchyma cell and a little red phloem fiber. Phloem parenchyma cells and sieve tube are similar on the cross section, and not easy to distinguish. Phloem ray is radial and composed of parenchyma cells arranged on the radial axis.

3.2.2 Cambium   It consists of flat rectangle cells arranged regularly and is internal to the secondary phloem. Cambium zone is composed of lines of cells on the cross section because it divides quickly.

3.2.3 Secondary xylem   It includes vessel, tracheid, xylem parenchyma and xylem fiber inside to the cambium. The vessels are the most recognizable, which are red colored, round or polygonal and arranged radially. Xylem ray consists of 1—2 layers of parenchyma cells, which are arranged radially in xylem and connected with phloem rays to form vascular rays and play a role of radial transportation.

Primary xylem lies in the secondary xylem and the center of root. Their vessels are small and round.

Observe cross section of root of *Panax ginseng* in the same way, its structure is characterized by resin canals in the secondary phloem and vessel of xylem are single row and radial loose arrangement, and have calcium oxalate cluster crystal in the phelloderm, xylem parenchyma cells and xylem ray.

4. Anomalous structure of root

4.1 Observe the tuber of *Polygonum multijflorum*. It has 4—11 similar rounds anomalous vascular bundle in the cortex on the light yellow-brown or red-brown cross section, named cloud brocade pattern. To distinguish periderm, parenchyma cells, abnormal vascular bundle and normal vascular bundle in the centre from outer to inner under the microscope. Cambium is round, anomalous vascular is complex, and a little is single vascular. The big normal collateral vascular bundle lies in the center of root. Primary xylem lies in the center.

4.2 Observe the root of *Achyranthes bidentata*. There are many small dots outside arranged in 2—4 rounds, that is, abnormal vascular bundle. Then observe the cross section under the microscope. The outside is phellem, which is composed of 4—8 layers of flat corked cells, and the inner side is several layers of parenchyma cells. The vascular tissue accounts for most of the root, and there are 2—4 rounds of open collateral abnormal vascular bundles outside. The normal vascular bundle is in the center of the root, and the primary xylem is diarch.

4.3 Observe the cross section of the old root of *Scutellaria baicalensis*. The xylem is located in the center of the root, and there are corked cell rings in the xylem.

Experiment report

1. Sketch the cross section of root of *Ranunculus japonicus* and *Stemona sessilifolia,* and note the name of every section.

2. Sketch the cross section of *Paeonia lactiflora*, and note the name of every section.

3. Sketch the cross section of *Achyranthes bidentala* or *Polygonum multiflorum*, and note the name of every section.

Questions

1. What are the differences of the primary structure between dicotyledon root and monocotyledon root?

2. What are the differences between the primary and the secondary structure of dicotyledon root? How is the secondary structure formed?

3. How is the abnormal structure of dicotyledon root formed?

实验十六 茎的构造

目的与要求

1. 掌握 双子叶植物茎的初生构造；双子叶植物木质茎与草质茎的次生构造；双子叶植物根状茎的异常构造。

2. 了解 裸子植物茎的次生构造。

实验材料

向日葵幼茎横切片，椴树茎横切片，薄荷茎横切片，大黄根茎横切片，松树茎横切片、径向纵切片、切向纵切片。

仪器用品

显微镜。

内容与方法

一、双子叶植物茎的初生构造

取向日葵幼茎横切片，置显微镜下观察，从外至内，依次可见以下结构。

1. 表皮 为位于最外层的一列扁平细胞，略有径向延长，排列紧密，偶见气孔。外壁常角质加厚，常见各种毛茸。

2. 皮层 为位于表皮层内侧的多层薄壁细胞，具细胞间隙。与根的初生构造相比，所占比例很小。靠近表皮的细胞较小，细胞内可见被染成绿色的类圆形叶绿体，细胞在角隅处增厚，形成厚角组织。其内为数层薄壁细胞，其中有小型分泌腔。皮层的最内一层细胞无凯氏带的分化。

3. 维管柱 位于皮层之内，所占面积宽广的细胞群，包括维管束、髓射线和髓。

（1）维管束 为数个大小不等的无限外韧型维管束构成。这些维管束排成一轮，由髓射线将束分开。每个维管束由初生韧皮部、束中形成层、初生木质部组成。

（2）初生韧皮部 位于维管束外方，其外侧有初生韧皮纤维。纤维横切面呈多角形，细胞壁明显加厚但未木质化，故染成绿色。在初生韧皮纤维内方是筛管、伴胞和韧皮薄壁细胞。

（3）初生木质部 位于维管束内方，包括原生木质部和后生木质部，导管横切面类圆形或多角形。根据导管分子直径大小和染色深浅可判断木质部类型：导管直径小，发生早，染色深，靠近茎中心的为原生木质部；导管直径大，发生晚，染色浅，接近束中形成层的为后生木质部。

（4）髓与髓射线 髓为位于茎的中央，也是维管柱中心的薄壁细胞群，排列疏松，常具有储存功能。由髓部放射的、排列着多束薄壁细胞组成髓射线，其内接髓部，外连皮层，具横向运输的功能。

二、双子叶植物茎的次生构造

（一）木质茎　取3~4年生椴树茎横切片，置显微镜下观察，从外至内，依次可见以下结构。

1. **周皮**　为位于最外几层排列整齐的扁方形细胞。周皮是取代表皮起次生保护作用的组织，由木栓层、木栓形成层和栓内层组成。

（1）木栓层　是在同一半径线上排列整齐的扁平细胞。细胞壁木栓质增厚，是死细胞。

（2）木栓形成层　由皮层细胞恢复分裂能力以后形成的，细胞呈扁平状，细胞质浓，可见细胞核。

（3）栓内层　一般由1~2层细胞组成，细胞质浓，染色较重。

2. **皮层**　为位于栓内层以内，多层类圆形或不规则的薄壁细胞组成。该层细胞不发达，较窄，偶见细胞内含草酸钙簇晶。

3. **韧皮部**　初生韧皮部常常被挤压成颓废组织，不易辨认。次生韧皮部细胞排列成梯形（底部靠近形成层），与排列成喇叭形的髓射线薄壁细胞相间分布。在切片中，明显可见染成红色的韧皮纤维与被染成绿色的韧皮薄壁细胞、筛管和伴胞呈横条状相间排列。

4. **形成层**　为位于韧皮部和木质部之间的数层分生组织，呈环状，由4~5层排列整齐的扁长细胞组成。

5. **木质部**　在形成层内方，在横切面上占有次生茎的最大体积，主要为次生木质部。由于其细胞直径的大小和细胞壁厚薄不同，可见数轮同心轮层，即年轮。注意观察早材和晚材在组织构造上的区别。紧靠髓部周围的一群小型导管即初生木质部，细胞少。

6. **髓**　位于茎的中央，由薄壁细胞组成。有的含草酸钙簇晶、单晶、黏液或单宁，所以部分细胞染色较深。

7. **髓射线**　由髓部薄壁细胞向外辐射状发出，直达皮层经木质部时，为1~2列细胞，至韧皮部时则扩大成喇叭状。

8. **维管射线**　在每个维管束之内，由木质部和韧皮部中的横向运输的薄壁细胞组成。位于木质部的称木射线，位于韧皮部的称韧皮射线。

（二）草质茎

取薄荷茎横切制片，置显微镜下观察，茎呈四方形，从外至内，依次可见以下结构。

1. **表皮**　为位于最外一层长方形细胞组成，外被角质层，有的可见腺毛、非腺毛或腺鳞。

2. **皮层**　为位于表层内数层排列疏松的薄壁细胞组成，较窄。在四个棱角内方，各有十余层厚角组织，其细胞在角隅处明显加厚，被染成红色。内皮层不明显。

3. **维管柱**

（1）维管束　由四个大的维管束（正对棱角）和其间2~4个较小维管束环状排列。韧皮部在外方，较窄，形成层成环，束间形成层明显。木质部在棱角处发达，导管单列，数行，纵向排列，在导管列之间为薄壁细胞组成的维管射线。

（2）髓与髓射线　髓部位于中央，发达，由大型薄壁细胞组成。髓射线为维管束间的薄壁细胞组成，宽窄不一。

此外，在薄荷茎的各部薄壁细胞中，有的还可见到扇形，具放射状纹理的橙皮苷结晶。

三、双子叶植物根状茎的异常构造

取大黄根茎横切片，置显微镜下观察，在低倍镜下可见木质部和宽广的髓部。髓部有多数星点状的周木式异型维管束。换高倍镜仔细观察其结构，形成层呈环状排列，内方为韧皮部，外方为木质部，射线呈星芒状射出。

四、裸子植物茎的次生构造

取3~4年生松树茎的三向切片，置显微镜下，分别观察其横切面、径向纵切面、切向纵切面。

图2　椴树茎横切面

图3　薄荷茎横切面

图4　大黄根茎横切面

1. 横切面 木质部无典型的木纤维，由管胞、木薄壁细胞和木射线组成。管胞呈四边形或六边形，具明显的细胞腔和木质化的断面。木射线呈辐射状条形，显示了其长度和宽度。可观察到明显的年轮。韧皮部由筛胞、韧皮薄壁细胞和韧皮射线组成。在皮层和维管柱中，分布有许多树脂道。

2. 径向纵切面 可见管胞呈长形，两端钝圆，纵向排列，其径向壁上有成行排列的呈两个同心圆状的具缘纹孔。射线细胞横向穿过管胞与纵轴垂直，细胞呈长方形，排成多列，显示了射线的长度和高度。

3. 切向纵切面 管胞呈菱形，纵向排列，所见射线是其横切面轮廓，呈纺锤状，显示了射线的高度、宽度、列数和两端细胞的形状。

实验报告

1. 绘制向日葵幼茎横切面结构简图，并注明各部分的名称。
2. 绘制椴树茎横切面结构简图，并注明各部分的名称。
3. 绘制薄荷茎横切面结构简图，并注明各部分的名称。
4. 绘制大黄根茎横切面结构简图，并注明各部分的名称。
5. 绘制松树茎横切面、径向纵切面和切向纵切面结构简图，注明各部位名称。

思考题

1. 双子叶植物茎与根的初生构造有何不同？
2. 双子叶植物茎的次生构造是怎样形成的？
3. 双子叶植物根状茎的异常构造有哪些特征？
4. 裸子植物茎与双子叶植物茎的次生构造有什么不同？

Experiment 16 — Structure of Stem I

Structure of Stem I

Aim and demand

1. Grasp the primary structure of dicotyledon stem, the secondary structure of dicotyledon stem (including woody and herbaceous stem), the anomalous structure of dicotyledon rhizome.

2. Understand the secondary structure of gymnosperm stem.

Experiment materials

Cross section of young stem of Xiangrikui (*Helianthus annuus* L.); cross section of stem of Duanshu (*Tilia amurensis* Rupr.), Bohe (*Mentha haplocalyx* Briq.), cross section of rhizome of Dahuang (*Rheum palmatum* L. or *Rheum tanguticum* Maxim. ex Balf. or *Rheum officinale* Baill.); Cross section, radial section and tangential section of Song (*Pinus tabulaeformis* Carr.).

Instrument and appliance

Microscope.

Contents and Procedures

1. Primary structure of dicotyledon stem

Take a cross section of the young stem of *Helianthus annuus* L. and observe it under the microscope from the outside to inwards, the following characters can be seen as follows.

1.1 Epidermis A row of flat cells are slightly elongated in a radial direction and closely arranged and located in the outermost layer with occasionally stomata; the outer walls are thick and cuticular, and a variety of trichomes are often present.

1.2 Cortex It is located on the inner side of the epidermis, constituted with multi-layer parenchyma cells. Compared with the primary structure of root, the proportion of cortex in stem is small. The cells near the epidermis are smaller and the similar round chloroplasts are dyed in green in the cells. The cells are thick at the corner and formed the collenchyma tissue. There are several layers of parenchyma cells inside the cortex and some of them are small secretory cavities. There's no casparian strip in the innermost part of the cortex.

1.3 Vascular cylinder It locates within the cortex and occupies a wide area of cell population, including vascular bundles, pith ray and pith.

1.3.1 Vascular bundle It consists of numbers of open collateral bundles with various sizes, and arranged in a round, and are separated by the pith rays. Each vascular bundle consists of primary phloem, fascicular cambiums and primary xylem.

1.3.2 Primary phloem   It is to located outside the vascular bundle and its outer are poly-angle primary phloem fibers. The fiber is obviously thickened, but not lignified, so dyed in green. Inside the fibers are sieve tubes, companion cells, and phloem parenchyma cells.

1.3.3 Primary xylem   It is located inside the vascular bundle, including protoxylem and metaxylem; the cross section of the vessel is round or polygonal. Xylem type can be judged by the diameter and color depth of the vessel. The protoxylem has characteristics of small diameter of vessel, early occurrence and deep staining and it is near the center of the stem; the metaxylem has large diameter of vessel, late occurrence, and thin staining and it is close to fascicular cambiums.

1.3.4 Pith and pith rays   Pith is located in the center of the stem, and it is parenchymatous, and the cells are loosely arranged in the center of vascular column. It has storage ability. Pith rays are composed of multiple parenchyma cells, which connect the pith and the cortex and have function of lateral transportation.

2. Secondary structure of dicotyledon stem

2.1 Woody stem   Take a cross section of 3-4 years old stem of *Helianthus annuus* and observe it from the outer to inner under the microscope, the following characters could be seen as follows.

2.1.1 Peridem   It is flat-rectangular arranged cells at the outermost layers and is the tissue that replaces epidermis for secondary protection, including phellem, phellogen and phelloderm.

2.1.1.1 Phellem   It consists of flat cells arranged regularly on the radius, and cell walls are thick suberization and cells are dead.

2.1.1.2 Phellogen   It comes from the cortical cells that restore division ability, and the cells are flat with dense cytoplasm and nucleus.

2.1.1.3 Phelloderm   It is 1-2 layers of cells with dense cytoplasm which is dyed deeply.

2.1.2 Cortex   It is composed of multi-layered round or irregular parenchyma cells located within the phelloderm. The cells are undeveloped and bit of narrow, and calcium oxalate clusters can be observed occasionally.

2.1.3 Phloem   The primary phloem has been damaged and not easy to distinguish. The cells of the secondary phloem are arranged in a trapezoid (near the cambium at the bottom), and are distributed with the horn-shaped parenchyma pith ray cells. In the section, it is obvious that the phloem fibers dyed in red and the phloem parenchyma cells, sieve tubes and companion cells dyed in green are arranged in a horizontal strip.

2.1.4 Cambium   It is the layer of meristem between the phloem and xylem and is ring-shaped, and consists of 4-5 layers of neatly arranged oblong cells.

2.1.5 Xylem   It is inside of the cambium and occupies the largest volume of secondary stem in the cross section, mainly consisting of secondary xylem. Due to the difference in cell diameter and cell wall thickness, several layers of concentric rings can be seen, that is, annual rings. Pay attention to the difference between early wood and late wood of tissue structure. There are a lot of small vessels near the pith, named primary xylem with few cells.

2.1.6 Pith   It lies on the cental stem, and is composed of parenchyma cells, some of which have calcium oxalate cluster crystal, slime or tannin and have resulted in the deep dyeing of these cells.

2.1.7 Pith rays   They are radially emitted by parenchyma cells of pith. They are 1-2 lines cells in the xylem and expand into a horn-shaped in the phloem.

2.1.8 Vascular rays   Every vascular ray composes of parenchyma cells of conducting ability on the crosswise in the xylem and phloem. The ones lie in the xylem named xylem rays and lie in the phloem named phloem rays.

2.2 Herbaceous stem Take a cross section of stem of *Mentha haplocalyx* and observe it under a microscope, then you will find the stem of it is square. From the outside inwards, the following characters can be seen as follows.

2.2.1 Epidermis It is composed of rectangular cells located in the outermost layer, which are covered by cuticles, and some are trichome (glandular hair, non-glandular hair or glandular scale).

2.2.2 Cortex It is narrow and composed of several parenchyma cells located within the epidermis. There are more than ten layers of collenchyma inside the four corners, and the cell corners are obviously thickened and dyed in red. The innermost layer is indistinct.

2.2.3 Vascular cylinder

2.2.3.1 Vascular bundle is composed four big vascular bundles (face to the four edges and corners) and 2-4 small vascular bundles among them. They arrange into round. Phloem on the outer and are narrow. Cambium is round and the interfascicular cambium is obvious. Xylem is developed on the edges and corners. Vessels are single line, and numbers of rows, arranged on the radial direction. Vascular rays composed of parenchyma cells lay among the vessel.

2.2.3.2 Pith and pith rays Pith is located in the center and is well developed, consisting of large, parenchymatous cells. Pith rays consist of parenchymatous cells between vascular bundles, which vary in width.

In addition, some parenchymatous cells of each part of stem are fan-shaped with radially textured hesperidin crystals.

3. Anomalous structure of dicotyledon rhizome

Take a cross section of the rhizome of *Rheum palmatum*, or *R. tanguticum* or *R. officinale*, and observe its xylem and broad pith under lower microscope. Many star-shaped anomalous vascular bundles are present in the pith, and each anomalous vascular bundle is amphivasal. Then turn to higher microscope and observe the structure carefully. The cambium is arranged as a ring, while phloem is at the inner and xylem is at the outer and the rays emit in a star shape.

4. Secondary structure of gymnosperm stem

Taking a cross section, radial section and tangential section of 3-4 years old stem of *Pinus tabulaeformis* and observe them separately under a microscope.

4.1 Cross section There is no vessel and typical xylem fiber in xylem which is mainly made of tracheid, parenchyma cells and xylem rays. The tracheids are quadrangular or hexagonal, with obvious cell cavities and lignified sections. The xylem rays are a radial bar, showing their length and width, and obvious annual ring can be observed. Phloem consists of sieve cells, parenchyma cells and phloem rays. There are lots of resin vessels in the cortex and vascular bundle.

4.2 Radial section It can be observed that the tracheids are elongated, bluntly rounded at both ends, and arranged longitudinally. The radial walls of them have two concentric circular pits arranged in rows. The ray cells pass through tracheids horizontally and are perpendicular to the vertical axis. The cells are rectangular and arranged in multiple rows, showing the length and height of the rays.

4.3 Tangential section The tracheids are prismatic and arranged longitudinally. The rays you can see their cross-sectional profile and are spindle-shaped, showing the height, width, number of columns, and the shape of the cells at both ends of the rays.

Experiment report

1. Draw the cross section structure of young stem of *Helianthus annuus*, and note the name of every section.

2. Draw the cross section structure of *Tilia amurensis*, and note the name of every section.

3. Draw the cross section structure of *Mentha haplocalyx*, and note the name of every section.

4. Draw the cross section structure of rhizome of *Rheum palmatum* or *R. tanguticum* or *R. officinale*, and note the name of every section.

5. Draw the cross section, radial section and tangential section structure of *Pinus tabulaeformis* and note the name of every section.

Questions

1. What are the differences between primary structure of dicotyledonous stem and root?

2. How is the secondary structure of the dicotyledonous stem formed?

3. What are the characteristics of anomalous structures of dicotyledonous rhizomes?

4. What are the differences between the secondary structure of gymnosperm stem and dicotyledon stem?

实验十七 茎的构造 II

目的要求

1. **掌握** 单子叶植物茎的构造。
2. **了解** 单子叶植物根状茎的构造。

实验材料

玉米茎的横切片；石斛茎的横切片；石菖蒲根茎的横切片。

仪器用品

显微镜。

内容与方法

（一）单子叶植物茎的构造

观察玉米茎的横切片，在低倍镜下由外向内区分表皮、基本组织和维管束，然后转换至高倍镜观察。

1. **表皮** 为茎的最外侧一层细胞，细胞扁长方形，排列整齐，外壁有较厚的角质层。

2. **基本组织** 为构成茎的主要部分。靠近表皮的数层细胞较小，排列紧密，胞壁增厚而木质化，形成厚壁组织，其内为薄壁组织。

3. **维管束** 呈散在分布。靠外侧的维管束小，内侧的渐大，没有皮层和髓之分。高倍镜下观察其中一个维管束，可见维管束外围有一圈由纤维组成的维管束鞘，里面为初生木质部和初生韧皮部，为有限外韧型。

初生木质部中的导管在横切面上排成"V"字形，上半部是后生木质部，含有一对并列的大导管，下半部为原生木质部，有 1~2 个纵向排列的小导管，少量薄壁细胞和一个大空腔。大空腔是由于茎的伸长而将环纹或螺纹导管扯破形成的。

初生韧皮部在初生木质部的外侧方，其中原生韧皮部已被挤压破坏，后生韧皮部明显，通常只含有筛管和伴胞。

观察石斛茎的横切面，其结构和玉米茎类似。

（二）单子叶植物根状茎的构造

观察石菖蒲根状茎横切片，在低倍镜下由外向内区分表皮、基本组织和维管束，然后转换至高倍镜观察。

1. **表皮** 为一层类方形排列紧密的细胞组成，外壁增厚且角质化。

2. **皮层** 由多层薄壁细胞组成，占较大部分，较宽广，其中散有油细胞、纤维束和叶迹维管

图 1 玉米茎横切面

图 2　石菖蒲茎横切面

束。维管束类圆形，周围细胞中含有草酸钙方晶，形成晶鞘纤维；叶迹维管束外韧型，周围有维管束鞘。内皮层明显，具有凯氏带。有的有根迹维管束通过。

3. 维管束　内皮层以内散有多数周木型维管束，靠内皮层排列紧密，有的为外韧型。维管束鞘纤维发达，周围细胞中含有草酸钙方晶。

实验报告

1. 比较单子叶植物茎和根茎结构的异同。
2. 绘制玉米茎的横切面简图。

思考题

1. 单子叶植物茎的基本结构有哪些？
2. 单子叶植物茎的维管束常见类型有哪些？

Experiment 17   Structure of Stem Ⅱ

Aim and demand

1. Grasp the structure of monocotyledon stem.
2. Understand the structure of monocotyledon rhizome.

Experiment materials

Cross section of stem Yumi (*Zea mays* L.) and Shihu (*Dendrobium nobile* Lindl.); cross section of rhizome of Shichangpu (*Acorus tatarinowii* Schott.).

Instruments and appliances

Microscope.

Contents and methods

1. The structure of monocotyledon stem

Observe the cross section of stem of *Zea mays* L., distinguish epidermis, ground tissue and vascular bundle under the low-power lens of microscope from the outer to inner, then turn to the high-power lens to observe.

1.1 Epidermis   It is composed of rectangle regularly parenchyma cells at the outer, and the outer walls have thicker cuticle.

1.2 Ground tissue   It is the main part of monocotyledonous plant. Layers of cells beneath the epidermis are arranged closely, and cell walls are thick and lignified, forming sclerenchyma. This is followed internally by parenchyma tissues.

1.3 Vascular bundle   They are dispersed into the ground tissue. Vascular bundles, without the border of cortex, pith and vascular cylinder, are smaller near the outer, and the larger near the inner. Turn to the higher microscope to observe a vascular bundle, a round vascular sheath composed of fiber out of the vascular bundle can be found. They only have primary xylem and phloem, and it was a closed collateral bundle.

Vessels of primary xylem are arranged in a "V" shape on the cross section. Metaxylem lies on the upper, and has two large vessels side by side. Protoxylem lies on the under with 1-2 radially arranged small vessels, a little parenchyma cells and a large cavity. The large cavity is formed by the elongation of the stem resulting in the rupture of the annular and spiral vessels.

Primary phloem lies in the outer of primary xylem, and protophloem has been already damaged. Metaphloem is obvious and has only a sieve tube and companion cell.

Observe cross section of stem of *Dendrobium nobile*, you may find similarities with stem of *Zea mays*.

2. The structure of monocotyledon rhizome

Observe the cross section of rhizome of *Acorus tatarinowii* to distinguish epidermis, cortex and vascular bundle under the low-power lens of microscope, then turn to the high-power lens to observe from the outer to inner.

2.1 Epidermis   It is composed of similar rectangle cells, and the outer walls are thicker and cuticle.

2.2 Cortex   It is composed of multilayer parenchyma cells, and occupies a larger part. The oil cell, fiber bundle and leaf trace and root trace vascular bundle are dispersed in the cortex. Vascular bundles are similarly rounded, and have calcium oxalate crystal among the cells, and formed the crystal fiber. Leaf trace vascular bundle is collateral and vascular sheath is to the outer of them. The endodermis, some of which have root trace vascular bundle, is distinct with a casparian strip.

2.3 Vascular bundle   A number of dispersed vascular bundles in the inner of the endodermis are termed amphivasal. They are arranged closed near the endodermis, and some of them are collateral. The fiber of vascular sheath is developed, and has calcium oxalate crystal among the cells.

💬 Experiment report

1. Compare the structures of stem and rhizome of monocotyledon.
2. Draw the cross section structure of *Zea mays*.

Questions

1. What are the basic structures of the stem of monocotyledon?
2. What are the common types of vascular bundle of monocotyledon stem?

实验十八 叶的内部构造

PPT

微课

目的要求

1. **掌握** 双子叶植物叶的内部构造。
2. **了解** 单子叶植物叶的内部构造。

实验材料

薄荷叶横切片，番泻叶横切片，淡竹叶横切片。

仪器用品

显微镜。

内容与方法

（一）双子叶植物叶片的构造

1. 观察薄荷叶的横切制片

（1）表皮　上表皮细胞长方形，下表皮细胞较小，均扁平，外被角质层，具气孔；表皮有腺鳞、腺毛和非腺毛。

（2）叶肉　栅栏组织为1列薄壁细胞，少有2列的，海绵组织为4~5列不规则且排列疏松的薄壁细胞组成。

（3）主脉　维管束外韧性，木质部位于主脉的近轴面（靠近上表皮），导管常2~6个纵列成数行，韧皮部位于木质部下方，较窄，细胞小，细胞呈多角形，形成层明显。主脉上、下表皮内侧有若干列厚角细胞。

本品表皮细胞、薄壁细胞和少数导管内有针簇状橙皮苷结晶。

图1 薄荷叶表皮

图2 薄荷叶横切面

2. 观察番泻叶的横切制片

（1）表皮　表皮细胞1列，外被角质层，上下表皮均有气孔及单细胞非腺毛；有的表皮细胞含黏液质，积聚于内壁。

（2）叶肉　叶肉为两面栅栏组织，均为1列细胞，上面的栅栏组织细胞较长，下面的较短；海绵组织2~3列细胞，细胞类圆形，有的含草酸钙簇晶。

（3）主脉　主脉上方有栅栏组织通过；维管束的上下均有微木化的纤维束，称为晶纤维。

图3 番泻叶横切面

（二）单子叶植物（禾本科）叶片的构造

观察淡竹叶横切制片，包括表皮、叶肉和主脉。

（1）表皮　上表皮主要为大型的运动细胞（泡状细胞）组成，呈扇形，细胞长方形，壁薄，径向延长；下表皮细胞较小，椭圆形，排列整齐，切向延长。上、下表皮均有角质层、气孔及单细胞非腺毛。

（2）叶肉　栅栏组织和海绵组织分化不明显。栅栏组织为一列短圆柱形的细胞，内含叶绿体并通过主脉；海绵组织由 1~3 列（多 2 列）排列较疏松的不规则圆形细胞组成。

（3）主脉　上部向下微凹，下部向外突起。中脉有一个较大型外韧型维管束，无形成层，四周有 1~2 列纤维包成维管束鞘，木质部导管稀少，排列成"V"型，其下部为韧皮部。在上、下表皮内侧有厚壁显微群。

💬 实验报告

1. 绘制薄荷叶横切面简图。
2. 绘制淡竹叶横切面简图。

🖊 思考题

比较双子叶植物和单子叶植物的叶在构造上的异同。

图 4　淡竹
叶横切面

习题

112

Experiment 18 | Internal Structure of Leaf

Aim and demand

1. Grasp the inner structure of dicotyledon leaves.
2. Know the inner structure of monocotyledon leaves.

Experiment materials

Cross section of Bohe (*Mentha halpocalyx* Briq.), cross section of Fanxieye (*Cassia acutifolia* Del.), cross section of Danzhuye (*Lophatherum gracile* Brongn.).

Instrument and appliance

Microscope.

Contents and Procedures

1. The structure of dicotyledon leaf

1.1 Observe cross section of leaf of *M. haplocalyx.*

1.1.1 Epidermis   Upper epidermis cell is rectangular, and lower epidermis cell is small and flat, cuticle, with stomata, The epidermis has glandular scales, glandular hair and no glandular hair.

1.1.2 Mesophyll   Palisade tissue is a layer of cells and spongy tissue is 4-5 layers of loose and parenchyma cells.

1.1.3 Vein   The vascular bundle is collateral, and the xylem lies on adaxial side (near upper epidermis). The vessels often split vertically into lines, with 2-6 vessels elements each line. The phloem cells under the xylem are narrower and smaller. The cambium of the vein is distinct. Collenchymatous tissues occur at the upper and lower layer of the main vein.

There are acicular and cluster crystals of hesperidinin in the epidermal cells, parenchyma cells and few vessels.

1.2 Observe cross section of leaf of *C. acutifolia.*

1.2.1 Epidermis   Epidermal cells are single row cells, covered with cuticle. There are stomata and unicellular non-glandular hairs on the both surfaces of the leaf. Some epidermal cells contain phlegm, accumulated in the inner wall.

1.2.2 Mesophyll   Palisade tissue exits on the both sides of the leaf and contains a row of cells. Palisade cells are longer on the adaxial surface and shorter on the abaxial surface. The spongy tissue contains 2-3 rows of round-like cells, and some of which have calcium oxalate cluster crystals.

1.2.3 Vein   Palisade tissue passes above the main vein. The vascular bundles are micro lignified fiber bundles and termed crystalline fibers.

2. The structure of monocotyledon leaves (Gramineae)

Observe cross section of the leaf of *L. gracile*

2.1 Epidermis Upper layer is mainly composed of large motor cells (vesicular cells), which are fan-shaped, similar rectangle, thin-walled and extended radial direction; the lower epidermis cells are small, elliptic, regularly arranged, and tangentially extended. Upper and lower epidermis are both covered by a cuticle layer, stoma, but no glandular hairs.

2.2 Mesophyll Palisade tissue and spongy tissue differentiation is indistinct. Palisade tissue consists of a row of short cylindrical cells with chloroplasts and passes through the main vein; Spongy tissue consists of loosely arranged irregular circular cells in 1-3 rows (more than 2 rows).

2.3 Vein The upper part is slightly concave downwards, and the lower part protrudes outwards. The midvein has a large external ductile vascular bundle without cambium and there are 1—2 rows of fibers surrounding the vascular bundle sheath. The vessels of the xylem, with the phloem under them, are few and arranged to 'V-shape'. There are sclerenchyma in the upper and lower layers of the vein.

Experiment report

1. Sketch the dissected structure of leaf of *M. haplocalyx*.
2. Sketch the dissected structure of leaf of *L. gracile*.

Questions

Compare structural similarities and differences of dicotyledon leaves and monocotyledon leaves.

实验十九　花、果实和种子的内部构造

PPT

微课

目标要求

1. 掌握　花、果实和种子的一般内部构造；被子植物花的胎座类型，区别有胚乳种子和无胚乳种子。

2. 了解　花粉粒的显微形态特征。

实验材料

柿蒂的横切片，月季花瓣横切片；鲜活植物的花：百合、豌豆、桔梗、石竹、黄瓜；松花粉、蒲黄、百合、芍药、南瓜等的花粉；枸杞果实的横切片；蓖麻、落花生种子，杏仁种子横切片。

仪器用品

显微镜、解剖镜、放大镜、解剖用具。

内容与方法

（一）花的内部构造

1. 萼片的内部构造　观察柿蒂的横切片，注意萼片和叶片的内部构造的相似点。观察柿的萼片表皮有无毛茸并判断其类型，上下表皮之间是否存在栅栏组织和海绵组织的分化，有无厚壁组织和草酸钙结晶。

2. 花瓣的内部构造　观察月季的花瓣制片，注意上表皮细胞的特征以及是否存在栅栏组织和海绵组织的分化。

3. 胎座的类型　取百合的子房作横切片，在放大镜或解剖镜下进行观察：可见子房壁内有3个子房室，每两个子房室之间的部分是2个心皮的结合处，是一隔膜，外侧有一凹陷，即腹缝线，其内有一堆维管束，是两个心皮侧束合并而成；在每个心皮中央有一中脉维管束（背束），与其相应的凹陷即背缝线、3个心皮内卷形成中轴，胚珠着生于腹缝线愈合形成的隔膜上，故称中轴胎座。

另取豌豆、桔梗、石竹、黄瓜等植物的子房作横切片，注意观察其子房的室数、心皮的数目及胚珠的着生情况，指出其各属于哪种类型的胎座。

4. 花粉粒的显微特征　分别选取1~2种花的花粉少许，用水合氯醛透化后，置于显微镜下观察。注意花粉形态、大小与类型，如单粒或复合；有无萌发孔，萌发孔的形状、位置与数目；表面光滑与否，或存在由小刺、瘤、颗粒等形成的各样雕纹特征。

医药大学堂
WWW.YIYAODXT.COM

115

（二）果实的内部构造

取枸杞果实横切片置于显微镜下观察下列部分。

1. **外果皮**　为一列扁平细胞，壁较薄，外被角质层，外缘作细齿状突起。

2. **中果皮**　为10余列薄壁细胞，外侧1~2列较小，中部细胞较大，有的细胞含草酸钙砂晶；维管束双韧型，散列。

3. **内果皮**　为一列椭圆形细胞，切向延长。

4. **种皮**　最外为一列石细胞，类长方形，侧壁及内壁呈"U"字形增厚。其下为3~4列被挤压的薄壁细胞。最内一层为扁平长方形薄壁细胞，微木质化。

5. **胚乳及胚根、子叶薄壁细胞**　含有脂肪油和颗粒状物。

（三）种子的内部构造

1. **有胚乳种子**　取蓖麻种子，小心剥去种皮，其内肥厚的部分为胚乳，用刀片平行于种子的宽面作纵切，把胚乳分为两半，用放大镜可观察到两片大而薄且叶脉清晰的子叶，同时可见胚根、极小的胚芽和很短的胚轴。

2. **无胚乳种子**　取花生种子，剥去种皮，其内为两片肥厚、乳白色、有光泽的子叶。打开两片子叶，可观察到胚芽是由一个主芽和两个侧芽组成。花生的胚轴位于胚芽下端，较粗壮，子叶着生于此。胚轴下方是胚根，突出于两片子叶之外，呈短喙状。

3. **杏仁横切片**

取杏仁横切片观察下列部位。

（1）种皮外表皮　为一列薄壁细胞组成，散生长圆形、卵形的橙黄色石细胞，上半部突出于表面，下半部埋在薄壁组织中。

（2）种皮内表皮　为一列薄壁细胞，含黄色物质。

（3）外胚乳　为数列颓废的薄壁组织。

（4）内胚乳　为一列长方形细胞，内含糊粉粒及脂肪油滴。

（5）子叶细胞　为多列多角形薄壁细胞，含糊粉粒，较大的糊粉粒中常有1细小草酸钙簇晶，并含脂肪油滴。

实验报告

1. 绘制柿蒂的萼片的结构简图，标明各部分的名称。
2. 绘制枸杞果实的横切片简图，标明各部分的名称。
3. 绘制蓖麻种子的剖面图，并注明各部分名称。

思考题

1. 花萼、花瓣和叶片的显微构造有何相似点？
2. 通过实验如何理解胚是一个幼小的植物体？

Experiment 19 | Internal Structure of Flower, Fruit and Seed

📋 Aim and demand

1. Grasp the structure of flower, fruit and seed.

2. Grasp the type of placenta of Angiospermae and distinguish the differences between albuminous seed and exalbuminous seed.

3. Understand the micromorphological characteristics of pollen grain.

⚫ Experiment materials

Cross section of sepal of Shidi (*Diospyros kaki* Thumb.) and petal of Yueji (*Rosa chinesis* Jacq.); flowers of fresh plants including Baihe (*Lilium brownie* F. E. Brown. var. *viridulum* Backer), Wandou (*Pisum sativum* L.), Jiegeng (*Platycodon grandiflorus* (Jacq.) A. DC.), Shizhu (*Dianthus chinensis* L.), Huanggua (Cucumis sativus Linn.); pollen of Yousong (*Pinus tabulaeformis* Carr.), Puhuang (*Typha angustifolia* L.), Baihe (*Lilium brownie* F. E. Brown. var. *viridulum* Backer), Shaoyao (*Paeonia lactiflora* Pall.) and Nangua (*Cucurbita moschata* (Duch.) Doiret.); cross section of fruit of Gouqi (*Lycium chinense* Mill.); cross section of seeds of Bima (*Ricinus communis* L.), Huasheng (*Arechis hypogaea* L.) and Xing (*Prunus armeniaca* L.).

⚙ Instrument and appliances

Microscope, stereomicroscope, magnifier and dissection tools.

🔍 Contents and Procedures

1. The structure of flower

1.1 The structure of calyx   Observe the cross section of sepal of *Diospyros kaki*, and note similarity of the internal structure of calyx and leaves. Observe the hairs on the epidermis and judge the type, and observe whether there is differentiation between the palisade tissue and the spongy tissue in the upper and lower epidermis. Furthermore, observe whether there are sclerenchyma and calcium oxalate crystals.

1.2 The structure of petal   Observe the cross section of flower of *Rosa chinensis* and pay attention to the characteristics of the upper epidermal cells and the presence of differentiation between the palisade tissue and the spongy tissue.

1.3 The type of placenta   Take a cross section of ovary of *Lilium pumilum* and observe it under magnifier or stereomicroscope. It is clear to see three ovary loculi, and the part between every two ovary loculi is the junction of two carpels. It is a septum with a depression on the outer side, that is, abdominal suture, in which there is a bunch of vascular bundles, is the combination of two carpel lateral bundles. In the center of each carpel, there is a midvein vascular bundle (dorsal bundle). The corresponding depression is the dorsal suture and three carpels are coiled to form the central axis. The ovules are atlach on the septum formed by the healing of the abdominal suture, therefore, it is called the central axial placenta.

Moreover, take cross sections of ovary of *Pisum sativum Platycodon grandiforum, Dianhus chinensis, Cueumis sativious,* to observe their loculi of ovary, numbers of carpels and located situation of ovary, and to determine the types of their placenta.

1.4 The microscopic characteristics of pollen grain　Take a little of pollen of 1-2 kinds of flowers, permeabilize with chloral hydrate and then place it under a microscope to observe. Pay attention to the shape, size and types of pollen grains. For instance, whether it is single or compound, and whether it has germinal aperture or not, and the shape, position, number of the germinal aperture. Observe the surface characteristics. Whether it is smooth or wave, and whether it has various carved features formed by small spines, nodules, particles, etc..

2．The structure of fruit

Observe the cross section of *Lycium chinense* under the microscope.

2.1 Exocarp　It is composed of a line of flat cells. Cell walls are thin, covered by a horny layer, and the outer edge is made of fine tooth-like protrusions.

2.2 Mesocarp　It consists of more than 10 lines of parenchynca cells. The outer 1~2 lines are smaller and the middle are bigger, some of which have sand of calcium oxalate crystals. Vascular bundles are bicollateral and have a dispensed arrangement.

2.3 Endocarp　It is composed of a line of oval cells, extending tangentially.

2.4 Seed coat　It is composed of a line of nearly-rectangle stone cells with "U-shaped" thickening of the lateral and inner wall. 'U-shaped' thickening. There are 3~4 lines crushed thin-walled cells under it. The innermost layer is composed of a flat rectangular thin-walled cells, with slightly lignified.

2.5 Endosperm, radicle and cotyledon parenchyma cells have fatty oil and granular substance.

3．The structure of seed

3.1 Endospermous seed　Take a seed of *Ricinus communis* and peel off the seed coat carefully. The thick part of it is endosperm. Cut the endosperm into two halves with a blade longitudinally in parallel with the wide surface of the seed. Then two large and thin leaves with clear cotyledons, and radicles, tiny germs and short hypocotyls can be observed by with a magnifer.

3.2 Non-endospermous seed　Take a seed of *Arachis hypogaea* and peel off the seed coat. There are two thick, milky, shiny cotyledons inside it. Separate the cotyledons, it can be observed that the germ is composed of a main bud and two lateral buds. The hypocotyl of *Amchis hypogee* is located at the lower end of the germ which is thick and cotyledone are born here. Below the hypocotyl is the radicle, projecting out of the two cotyledons and resembling a short beak.

3.3 The structure of seed

Take a cross section of *Pinus armeniaca* to observe as follows.

3.3.1 Outer epidermis of seed coat　It is composed of a line of thin-walled cells, and scattered with round, oval orange-yellow stone cells. The upper part protrudes from the surface, and the lower half are buried in the parenchyma.

3.3.2 Inter epidermis of seed coat　It is composed of a line of thin-walled cells, containing yellow substances.

3.3.3 Perisperm　It is composed of several lines of decadent parenchyma.

3.3.4 Endosperm   It is composed of a line of rectangle cells, and contains aleurone grain and fatty oil.

3.3.5 Cotyledon   It consists of lines of thin-walled cells, containing aleurone granules which the bigger ones often contain small calcium oxalate crystal and fatty oil droplets.

Experiment report

1. Draw the cross section structure of *Diospyros kaki*, and note the name of every section.

2. Draw the cross section structure of fruit of *Lycium chinense*, and note the name of every section.

3. Draw the cross section structure of seed of *Prunus armeniaca*, and note the name of every section.

Questions

1. What are the similarities among micro structures of calyx, petal and leaf?

2. How to understand that an embryo is a young plant through experiment?

附　录

附录一　实验室管理规则

1. 遵守实验安排，不得迟到、早退或无故缺席。

2. 听从指导教师和管理人员的安排，对号入座。

3. 课前应备好并携带实验指导、教材等参考资料及植物解剖器具等。

4. 遵守课堂纪律。需身着白大褂进入实验室。严禁在实验过程中做与实验无关的事情。

5. 爱护标本、样品等公共教具。实验完成后材料必须归还，不得私自携带出室外（如有破损，按价赔偿）。

6. 正确操作显微镜等实验仪器。如发现有损坏，应及时主动报告。实验结束后关闭电源，将物品复原。

7. 具有腐蚀性的液体必须倒在废液缸内，不得倒入下水道中。废物应放入垃圾桶中。

8. 实验结束时，应将个人物品清洗、收好，台面周边清理干净后方可离开。值日同学负责清扫实验室卫生，检查水、电及门窗，保证安全、整洁。

9. 实验前应提前预习，实验时必须按照老师指导的方法操作，实验结束应按时完成实验报告。

Appendix 1　Regulations of laboratory

1. Comply with the experiment schedule, do not be late, leave early or be absent without reason.

2. Follow the instructions of the instructor and the management, and take the right seats.

3. Prepare and carry the experimental guidance, teaching materials and other reference tools such as plant dissectors before the class.

4. Observe class room disciplines. It is necessary to wear lab coat when entering the laboratory. Anything irrelevant to the experiment is strictly forbidden.

5. Take good care of samples and other public teaching aids. The materials must be returned after the completion of the experiment, and shall not be taken out of the laboratory without permission (in case of damage, compensation shall be made according to the price).

6. Operate the microscope and other experimental instruments correctly. If any damage is found, report it promptly. After the experiment, turn off the power and restore the objects.

7. The corrosive liquid must be poured into the waste tank, not into the sewer. The waste should be put into the garbage can.

8. At the end of experiment, personal belongings should be cleaned and put away, and the table must be cleaned before leaving. Students on duty are responsible for cleaning and inspection to ensure tidiness and safety of the laboratory.

9. You are supposed to preview the experiment content in advance, and operate the experiment correctly according to the method guided by the teacher during the experiment, and finish the experiment report on time.

附录二　常用试剂的配制

1. **镜头清洁剂**　取乙醚 7ml 和无水乙醇 3ml，混合，装入滴瓶备用，用于擦拭显微镜镜头上的油迹和污垢等（塞紧瓶口，以免挥发）。

2. **稀甘油**　取甘油 30ml，加蒸馏水稀释成 100ml 即得。稀甘油能使细胞稍透明及溶解某些水溶性细胞后含物，并使材料保持湿润和软化，常与水合氯醛试剂同用作临时封藏剂，可防止水合氯醛晶体析出。

3. **水合氯醛试剂**　取水合氯醛 50g，加蒸馏水 15ml，加入甘油 10ml 使其溶解，即得。本试液能迅速透入组织，使干燥而收缩的细胞膨胀，细胞组织透明清晰，并能溶解淀粉粒、树脂、蛋白质和挥发油等。

4. **间苯三酚试液**　取间苯三酚 1g，加 90% 乙醇 5ml 溶解后，加甘油 5ml，摇匀，即得。用于鉴别木质化细胞壁。

5. **稀碘液**　取碘化钾 1g 溶于 100ml 蒸馏水中，再加碘 0.3g，储于棕色瓶中。稀碘液可使淀粉粒显蓝色，糊粉粒显黄色。

6. **苏丹Ⅲ试液**　取苏丹Ⅲ 0.01g，加 90% 乙醇 5ml 溶解后，加甘油 5ml，摇匀，储于棕色玻璃瓶中，保存期为两个月。

7. **钌红试液**　取 10% 乙酸钠溶液 1~2ml，加钌红适量使成酒红色，即得。本试液应临用时新制，可使黏液显红色。

8. **α-萘酚试液**　取 α-萘酚 1.5g 溶于 95% 乙醇 10ml，即得。应用时滴加本试液，1~2min 后再滴加 80% 硫酸 2 滴，可使菊糖显紫色。

9. **氯化锌碘试液**　取氯化锌 20g，溶于 85ml 蒸馏水后，滴加碘的碘化钾溶液（碘化钾 3g，碘 1.5g，水 60ml），不断振摇至饱和，至没有碘的沉淀出现为止，至棕色瓶内保存。

10. **番红染液**　番红是一种碱性染料，可使木质化、木栓化、角质化的细胞壁及细胞核中的染色质和染色体染成红色。在植物组织制片中常与固绿配染。常用配方有下列两种。

（1）番红水液　取番红 0.1g，溶于 100ml 蒸馏水中，过滤后，即得。

（2）番红酒液　取番红 0.5g，溶于 50% 乙醇 100ml 中，过滤后，即得。

11. **固绿染液**　取固绿 0.1g，溶于 95% 乙醇 100ml 中，过滤后使用。

12. **F.A.A 固定液（又称万能固定剂）**　福尔马林（38% 甲醛）5ml，冰醋酸 5ml，70% 乙醇 90ml。幼嫩材料用 50% 乙醇代替 70% 乙醇，可防止材料收缩；还可以加入甘油 5ml，以防止蒸发和材料变硬。此液兼可作为保存剂使用。

13. **洗液**　取工业用重铬酸钾 8~10g 于 10ml 清水中，加热使溶解，待冷却后，慢慢加入工业用浓硫酸 100ml，即得。储藏于玻璃容器中。

Appendix 2　The preparation and use of frequently used reagents

1. Cleaner reagent of lens　7ml ethyl ether and 3ml anhydrous ethanol were taken, mixed and put into a drop bottle for later use. This detergent is used to clean the oil and dirt on the lens of the microscope. (Note: the bottle must be corked tightly in order to prevent the detergent from volatilising.)

2. Thin glycerol　Take glycerine 30ml and dilute to 100ml with distilled water. The thin glycerine could make cells slightly transparent and dissolve some hydrosoluble ergastic substance, and make the material keep wet and soften. The glycerol is usually used with chloral hydrate as a temporary sealing reagent and can prevent chloral hydrate crystal precipitation.

3. Chloral hydrate reagent　Take chloral hydrate 50g mixed homogeneously with distilled water 15ml and glycerine 10ml. The reagent can seep into the material quickly, make the dry and contracted cells swell, the tissue of cells transparent and distinct. Besides, the reagent can dissolve substances such as starch grains, resin, protein, and essential oil and so on.

4. Phloroglucinol solution　Take phlorogucinol 1g dissolved in alcohol solution (90%) 5ml, then mixed homogeneously with glycerol 5ml. The solution is used to identify lignified cell wall.

5. Thin iodine solution　Take potassium iodide 1g dissolved in distilled water 100ml, then add iodine 0.3g to the solution. This reagent can make the starch grain turn blue, and the alcurone grain turn yellow. Please preserve the solution in a brown bottle.

6. Sudan Ⅲ reagent　Take Sudan Ⅲ 0.01g dissolved in alcohol solution (90%) 5ml, and mixed homogeneously with glycerine 5ml. Please preserve the solution in a brown bottle for two months..

7. Ruthwnium red reagent　Take solution (10%) of sodium acetic 1-2ml. Add an appropriate amount of Ruthwnium red to make wine red. The reagent should be prepared only to make the mucus red.

8. α-naphthol reagent　Take α-naphthol 1.5g dissolved in alcohol solution (95%) 10ml. When applied, add the test solution, and then add 2 drops of 80% sulfuric acid after 1-2min to make inulin purple.

9. $ZnCl_3$-I reagent　Take Zine chloride 20g dissolved in distilled water 85ml, then add Iodine's solution of Potassium Iodine (KI: 3g, I_2: 1.5g, distilled water: 60ml). Shake the solution in a bottle until it is saturate and there are no sediment of I_2. Preserve the solution in brown bottle.

10. Safranine dye solution　Saffron is an alkaline dye which can stair lignified, corked, keratinized cell walls and chromatin and chromosomes in the nucleus red. It is often dyed with green-fixed dye solution in plant tissue preparation. The two commonly used formulas are as follows.

（1）Safranine dissolved in water　Take safranine 0.1g dissolved in distilld water 100ml, then get the solution filtered.

（2）Safranine dissolved in alcohol　Take safranine 0.5g dissolved in alcohol solution (50%) 100ml, then get the solution filtered.

11. Green-fixed dye solution　Take green-fixed dye 0.1g dissolved in 95% ethanol 100ml,, then get the solution filtered.

12. F. A. A reagent for fixing (also called universal reagent for fixing)　Take formaldehyde solution

（38%）5ml mixed with glacial acetic acid 5ml and ethanol solution（70%）90ml. If the material is rather tender, please use ethanol 50% in stead of ethanol 70% in order to protect the material from shrinking. Besides, 5ml glycerol can also be added to prevent evaporation and material hardening. The reagent also can be used for preserving.

13. Chromicacid lotion   Take 8-10g of $K_2Cr_2O_7$ for industrial use in 10ml of water, heat it to be dissolved. After cooling, slowly add 100ml of the Industrial concentrated sulfuric. Please preserve the lotion in a glass container.

参考文献

［1］Cronquist A. An Integrated System of Classification of Flowering Plants. New York: Columbia University Press, 1981.

［2］Davis PH and VH Heywood. Principles of Angiosperm Taxonomy. London: Printed in Great British By T. and A. Constable Led, 1963.

［3］George HM Lawrence. Taxonomy of Vascular Plants. New York: Macmillan, Company, 1965.

［4］James G, Harris&Melinda Woolf Harris. Plant Identification Terminology: An Illustrated Glossary. Payson：Spring Lake Publishing, 1994.

［5］Stem et al., Introductory Plant Biology, Ed. Nine McGraw-Hill Higher Education, New York: U. S. A 2003.

［6］Weier TE, Barbour MG, Sticking CR et al. Botany, An Introduction To Plant Biology. California: John Wiley & Sons, Inc., 1982.

［7］福斯特（Foster AS），小吉福德（Gifford EMJr.）著．李正理，张新英，李荣敖等译．维管植物比较形态学．北京：科学出版社，1983.

［8］刘春生．药用植物学．北京：中国中医药出版社，2016.

［9］路金才．药用植物学．北京：中国中医药出版社，2016.

［10］谭献河，王冰．药用植物学实验指导．北京：中国中医药出版社，2018.

［11］王旭红．药用植物学实验指导．北京：中国医药科技出版社，2018.

［12］中国科学院《中国植物志》编辑委员会．中国植物志．北京：科学出版社，1959-2004.

［13］中国科学院植物研究所．中国高等植物图鉴（1-5 册，补编1-2 册）．北京：科学出版社，1972-1989。